Higher

Chemistry

2004 Exam

2005 Exam

2006 Exam

2007 Exam

2008 Exam

Leckie × Leckie

First exam published in 2004.

Published by Leckie & Leckie Ltd, 3rd Floor, 4 Queen Street, Edinburgh EH2 1JE

tel: 0131 220 6831 fax: 0131 225 9987 enquiries@leckieandleckie.co.uk www.leckieandleckie.co.uk

ISBN 978-1-84372-673-9

A CIP Catalogue record for this book is available from the British Library.

Leckie & Leckie is a division of Huveaux plc.

Leckie & Leckie is grateful to the copyright holders, as credited at the back of the book, for permission to use their material.
Every effort has been made to trace the copyright holders and to obtain their permission for the use of copyright material.
Leckie & Leckie will gladly receive information enabling them to rectify any error or omission in subsequent editions.

[BLANK]

FOR OFFICIAL USE

Total
Section B

X012/301

NATIONAL
QUALIFICATIONS
2004

WEDNESDAY, 2 JUNE
9.00 AM – 11.30 AM

CHEMISTRY
HIGHER

Fill in these boxes and read what is printed below.

Full name of centre

Town

Forename(s)

Surname

Date of birth
Day Month Year   Scottish candidate number   Number of seat

Reference may be made to the Chemistry Higher and Advanced Higher Data Booklet (1999 edition).

SECTION A—Questions 1–40

Instructions for completion of **Section A** are given on page two.

SECTION B

1   All questions should be attempted.

2   The questions may be answered in any order but all answers are to be written in the spaces provided in this answer book, and must be written clearly and legibly in ink.

3   Rough work, if any should be necessary, should be written in this book and then scored through when the fair copy has been written.

4   Additional space for answers and rough work will be found at the end of the book. If further space is required, supplementary sheets may be obtained from the invigilator and should be inserted inside the **front** cover of this book.

5   The size of the space provided for an answer should not be taken as an indication of how much to write. It is not necessary to use all the space.

6   Before leaving the examination room you must give this book to the invigilator. If you do not, you may lose all the marks for this paper.

SCOTTISH
QUALIFICATIONS
AUTHORITY

SECTION A

1. Check that the answer sheet provided is for Chemistry Higher (Section A).

2. Fill in the details required on the answer sheet.

3. **In questions 1 to 40 of the paper, an answer is given by indicating the choice A, B, C or D by a stroke made in INK in the appropriate place in the answer sheet—see the sample question below.**

4. **For each question there is only ONE correct answer.**

5. Rough working, if required, should be done only on this question paper, or on the rough working sheet provided—**not** on the answer sheet.

6. At the end of the examination the answer sheet for Section A **must** be placed **inside** the front cover of this answer book.

This part of the paper is worth 40 marks.

SAMPLE QUESTION

To show that the ink in a ball-pen consists of a mixture of dyes, the method of separation would be

        A fractional distillation

        B chromatography

        C fractional crystallisation

        D filtration.

The correct answer is B—chromatography. A **heavy** vertical line should be drawn joining the two dots in the appropriate box in the column headed **B** as shown **in the example on the answer sheet**.

If, after you have recorded your answer, you decide that you have made an error and wish to make a change, you should cancel the original answer and put a vertical stroke in the box you now consider to be correct. Thus, if you want to change an answer **D** to an answer **B**, your answer sheet would look like this:

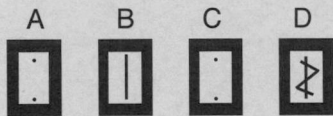

If you want to change back to an answer which has already been scored out, you should **enter a tick (✓)** to the RIGHT of the box of your choice, thus:

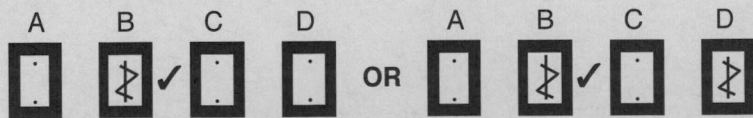

1. Which of the following solids has a low melting point and a high electrical conductivity?

   A   Iodine

   B   Potassium

   C   Silicon oxide

   D   Potassium fluoride

2. Two experiments are set up to study the corrosion of an iron nail.

   Experiment 1   Experiment 2

   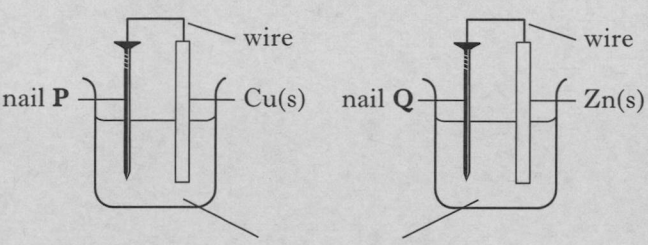

   NaCl(aq) containing ferroxyl indicator

   After a short time, a blue colour will have appeared at

   A   both **P** and **Q**

   B   neither **P** nor **Q**

   C   **P** but not at **Q**

   D   **Q** but not at **P**.

3. In which of the following compounds do **both** ions have the same electron arrangement as argon?

   A   Calcium sulphide

   B   Magnesium oxide

   C   Sodium sulphide

   D   Calcium bromide

4. What volume of sodium hydroxide solution, concentration $0·4\ mol\,l^{-1}$, is needed to neutralise $50\ cm^3$ of sulphuric acid, concentration $0·1\ mol\,l^{-1}$?

   A $25\ cm^3$

   B $50\ cm^3$

   C $100\ cm^3$

   D $200\ cm^3$

5. Like atoms, molecules can lose electrons to form positive ions.

   1. $[^1H_2{}^{16}O]^+$ 2. $[^1H_2{}^{17}O]^+$ 3. $[^1H_2{}^{18}O]^+$

   4. $[^2H_2{}^{16}O]^+$ 5. $[^2H_2{}^{17}O]^+$ 6. $[^2H_2{}^{18}O]^+$

   Which of the following pairs has ions of the same mass?

   A   1 and 4

   B   2 and 5

   C   3 and 6

   D   3 and 4

6. Which of the following graphs could represent the change in the rate of a reaction between magnesium ribbon and hydrochloric acid?

   A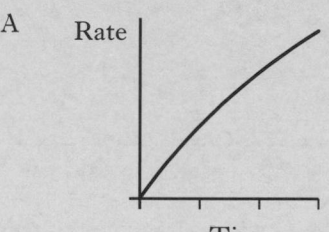

   B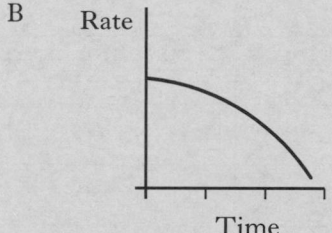

   C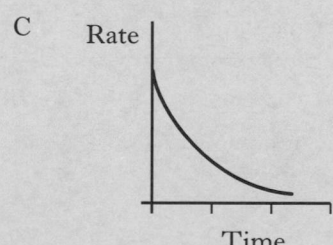

   D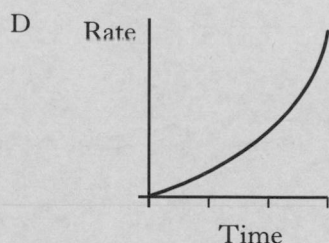

[Turn over

7. 1 mol of hydrogen gas and 1 mol of iodine vapour were mixed and allowed to react. After t seconds, 0·8 mol of hydrogen remained.

The number of moles of hydrogen iodide formed at t seconds was

A 0·2

B 0·4

C 0·8

D 1·6.

8.

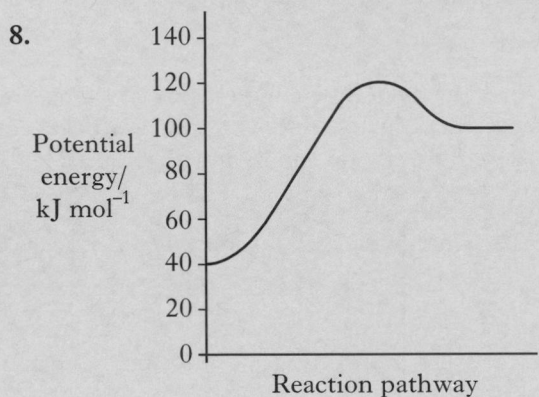

Reaction pathway

Which of the following sets of data applies to the reaction represented by the above energy diagram?

	Enthalpy change	Activation energy/ kJ mol^{-1}
A	Exothermic	60
B	Exothermic	80
C	Endothermic	60
D	Endothermic	80

9. Which of the following elements has the greatest electronegativity?

A Caesium

B Oxygen

C Fluorine

D Iodine

10. As the relative atomic mass in the halogens increases

A the boiling point increases

B the density decreases

C the first ionisation energy increases

D the atomic size decreases.

11. Which of the following elements would require the most energy to convert one mole of gaseous atoms into gaseous ions each carrying two positive charges?

(You may wish to use the data booklet.)

A Scandium

B Titanium

C Vanadium

D Chromium

12. Which of the following compounds has polar molecules?

A CH_4

B CO_2

C NH_3

D CCl_4

13. $2NO(g) + O_2(g) \rightarrow 2NO_2(g)$

How many litres of nitrogen dioxide gas could theoretically be obtained in the reaction of 1 litre of nitrogen monoxide gas with 2 litres of oxygen gas?

(All volumes are measured under the same conditions of temperature and pressure.)

A 1

B 2

C 3

D 4

14. Which of the following gases has the same volume as 128·2 g of sulphur dioxide gas?

(All volumes are measured under the same conditions of temperature and pressure.)

A   2·0 g of hydrogen

B   8·0 g of helium

C   32·0 g of oxygen

D   80·8 g of neon

15. 5 g of copper is added to excess silver(I) nitrate solution. After some time, the solid present is filtered off from the copper(II) nitrate solution, washed with water, dried, and weighed.

The final mass of the solid will be

A less than 5 g

B 5 g

C 10 g

D more than 10 g.

16. Which of the following equations represents a reaction which takes place during reforming?

A $C_6H_{14} \rightarrow C_6H_6 + 4H_2$

B $C_4H_8 + H_2 \rightarrow C_4H_{10}$

C $C_2H_5OH \rightarrow C_2H_4 + H_2O$

D $C_8H_{18} \rightarrow C_4H_{10} + C_4H_8$

17. Which of the following is a ketone?

A

$$\overset{O}{\overset{\|}{C}} - H$$

B

$$\overset{O}{\overset{\|}{C}} - CH_3$$

C

$$\overset{O}{\overset{\|}{C}} - O - H$$

D

$$\overset{O}{\overset{\|}{C}} - O - CH_3$$

18. Which of the following is an isomer of 2,2-dimethylpentan-1-ol?

A $CH_3CH_2CH_2CH(CH_3)CH_2OH$

B $(CH_3)_3CCH(CH_3)CH_2OH$

C $CH_3CH_2CH_2CH_2CH_2CH_2CH_2CH_2OH$

D $(CH_3)_2CHC(CH_3)_2CH_2CH_2OH$

19. Ethene is used in the manufacture of addition polymers.

What type of reaction is used to produce ethene from ethane?

A Addition

B Cracking

C Hydrogenation

D Oxidation

20. The compound $CH_3CH_2COO^-Na^+$ is formed by reaction between sodium hydroxide and

A propanoic acid

B propan-1-ol

C propene

D propanal.

21. Which of the following is **not** a correct statement about methanol?

A It is a primary alkanol.

B It can be oxidised to methanal.

C It can be made from synthesis gas.

D It can be dehydrated to an alkene.

22. Ammonia is manufactured from hydrogen and nitrogen by the Haber Process

$$3H_2 \quad + \quad N_2 \quad \rightleftharpoons \quad 2NH_3$$

If 80 kg of ammonia is produced from 60 kg of hydrogen, what is the percentage yield?

A $\dfrac{80}{340} \times 100$

B $\dfrac{80}{170} \times 100$

C $\dfrac{30}{80} \times 100$

D $\dfrac{60}{80} \times 100$

23. What mixture of gases is known as synthesis gas?

A Methane and oxygen

B Carbon monoxide and oxygen

C Carbon dioxide and hydrogen

D Carbon monoxide and hydrogen

24. Part of a polymer chain is shown below.

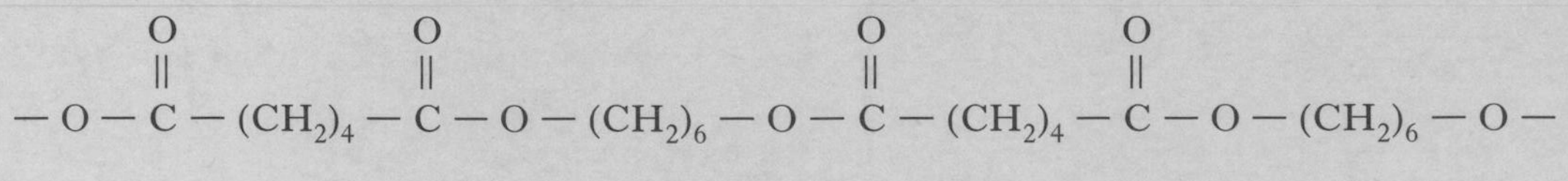

$$-O-\overset{\overset{O}{\|}}{C}-(CH_2)_4-\overset{\overset{O}{\|}}{C}-O-(CH_2)_6-O-\overset{\overset{O}{\|}}{C}-(CH_2)_4-\overset{\overset{O}{\|}}{C}-O-(CH_2)_6-O-$$

Which of the following compounds, when added to the reactants during polymerisation, would stop the polymer chain from getting too long?

A $HO-\overset{\overset{O}{\|}}{C}-(CH_2)_4-\overset{\overset{O}{\|}}{C}-OH$

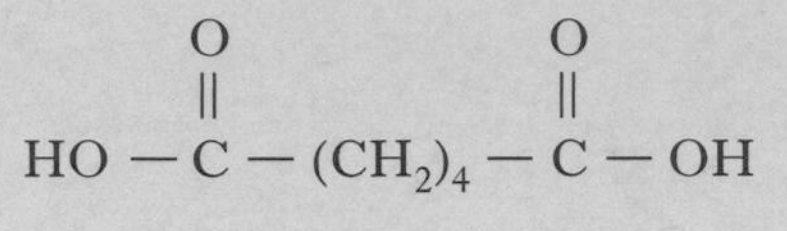

B $HO-(CH_2)_6-OH$

C $HO-(CH_2)_5-\overset{\overset{O}{\|}}{C}-OH$

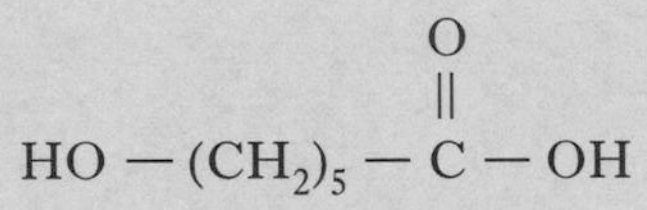

D CH_3-OH

---

25. Which of the following polymers is used in making bullet-proof vests?

A Kevlar

B Biopol

C Poly(ethenol)

D Poly(ethyne)

26. Which of the following is a structural formula for glycerol?

A CH$_2$OH
   |
  CHOH
   |
  CH$_2$OH

B CH$_2$OH
   |
  CH$_2$
   |
  CH$_2$OH

C CH$_2$OH
   |
  CH$_2$OH

D CH$_2$OH
   |
  CHOH
   |
  CH$_2$COOH

27. Fats have higher melting points than oils because comparing fats and oils

A fats have more hydrogen bonds

B fat molecules are more saturated

C fat molecules are more loosely packed

D fats have more cross-links between molecules.

28. The monomer units used to construct enzyme molecules are

A alcohols

B esters

C amino acids

D fatty acids.

29. Which of the following compounds is a raw material in the chemical industry?

A Ethene

B Ammonia

C Sulphuric acid

D Sodium chloride

30. Consider the reaction pathway shown.

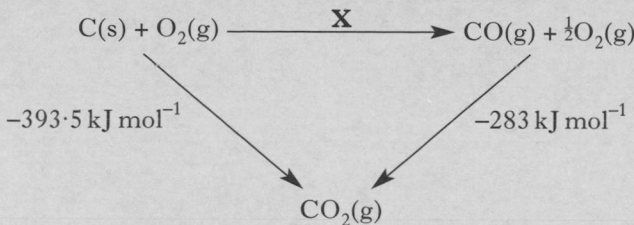

$$C(s) + O_2(g) \xrightarrow{\quad X \quad} CO(g) + \tfrac{1}{2}O_2(g)$$

$-393 \cdot 5 \, kJ \, mol^{-1}$ $\qquad\qquad -283 \, kJ \, mol^{-1}$

$$CO_2(g)$$

According to Hess's Law, what is the enthalpy change for reaction **X**?

A $\quad +110 \cdot 5 \, kJ \, mol^{-1}$

B $\quad -110 \cdot 5 \, kJ \, mol^{-1}$

C $\quad -676 \cdot 5 \, kJ \, mol^{-1}$

D $\quad +676 \cdot 5 \, kJ \, mol^{-1}$

31. Which line in the table applies correctly to the use of a catalyst in a chemical reaction?

	Position of equilibrium	Effect on value of ΔH
A	Moved to right	Decreased
B	Unaffected	Increased
C	Moved to left	Unaffected
D	Unaffected	Unaffected

32. Some solid ammonium chloride is added to a dilute solution of ammonia.

Which of the following ions will decrease in concentration as a result?

A Ammonium

B Hydrogen

C Hydroxide

D Chloride

33. The pH of a solution of hydrochloric acid was found to be $2 \cdot 5$.

The concentration of the $H^+(aq)$ ions in the acid must be

A $\quad$ greater than $0 \cdot 1 \, mol \, l^{-1}$

B $\quad$ between $0 \cdot 1$ and $0 \cdot 01 \, mol \, l^{-1}$

C $\quad$ between $0 \cdot 01$ and $0 \cdot 001 \, mol \, l^{-1}$

D $\quad$ less than $0 \cdot 001 \, mol \, l^{-1}$.

34. Which of the following is the best description of a $0 \cdot 1 \, mol \, l^{-1}$ solution of sulphuric acid?

A Dilute solution of a strong acid

B Dilute solution of a weak acid

C Concentrated solution of a strong acid

D Concentrated solution of a weak acid

35. Excess marble chips (calcium carbonate) were added to $100 \, cm^3$ of $1 \, mol \, l^{-1}$ hydrochloric acid. The experiment was repeated using the same mass of the marble chips and $100 \, cm^3$ of $1 \, mol \, l^{-1}$ ethanoic acid.

Which of the following would have been the same for both experiments?

A The time taken for the reaction to be completed

B The rate at which the first $10 \, cm^3$ of gas was evolved

C The mass of marble chips left over when the reaction had stopped

D The average rate of the reaction

36. Which line in the table is correct for $0 \cdot 1 \, mol \, l^{-1}$ sodium hydroxide compared with $0 \cdot 1 \, mol \, l^{-1}$ aqueous ammonia?

	pH	Conductivity
A	higher	lower
B	higher	higher
C	lower	higher
D	lower	lower

37. During a redox process in acid solution, iodate ions, $IO_3^-(aq)$, are converted into iodine, $I_2(aq)$.

$$IO_3^-(aq) \quad \rightarrow \quad I_2(aq)$$

The numbers of $H^+(aq)$ and $H_2O(\ell)$ required to balance the ion-electron equation for the formation of 1 mol of $I_2(aq)$ are, respectively

A 3 and 6

B 6 and 3

C 6 and 12

D 12 and 6.

38. Ammonia reacts with magnesium as shown.

$$3Mg(s) + 2NH_3(g) \rightarrow (Mg^{2+})_3(N^{3-})_2(s) + 3H_2(g)$$

In this reaction, ammonia is acting as

A   an acid

B   a base

C   an oxidising agent

D   a reducing agent.

39. Induced nuclear reactions can be described in a shortened form

$$T\,(x, y)\,P$$

where the participants are the target nucleus (T), the bombarding particle (x), the ejected particle (y) and the product nucleus (P).

Which of the following nuclear reactions would **not** give the product nucleus indicated?

A $^{14}_{7}N$ (α, p) $^{17}_{8}O$

B $^{236}_{93}Np$ (p, α) $^{238}_{92}U$

C $^{10}_{5}B$ (α, n) $^{13}_{7}N$

D $^{242}_{96}Cf$ (n, α) $^{239}_{94}Pu$

40. Which of the following equations represents a nuclear fission process?

A $^{40}_{19}K$ + $^{0}_{-1}e$ $\rightarrow$ $^{40}_{18}Ar$

B $^{2}_{1}H$ + $^{3}_{1}H$ $\rightarrow$ $^{4}_{2}He$ + $^{1}_{0}n$

C $^{235}_{92}U$ + $^{1}_{0}n$ $\rightarrow$ $^{90}_{38}Sr$ + $^{144}_{54}Xe$ + $2^{1}_{0}n$

D $^{14}_{7}N$ + $^{1}_{0}n$ $\rightarrow$ $^{14}_{6}C$ + $^{1}_{1}p$

Candidates are reminded that the answer sheet MUST be returned INSIDE the front cover of this answer book.

[Turn over for SECTION B on *Page ten*

SECTION B

Marks

1. (*a*) Complete the table below by adding the name of an element from **elements 1 to 20** of the Periodic Table for each of the types of bonding and structure described.

Bonding and structure at room temperature and pressure	Name of element
metallic solid	sodium
monatomic gas	
covalent network solid	
discrete covalent molecular gas	
discrete covalent molecular solid	

2

(*b*) Why do metallic solids such as sodium conduct electricity?

1

(3)

DO NOT WRITE IN THIS MARGIN

Marks

2. Phosphorus-32 is a radioisotope that decays by beta-emission.

(*a*) Write the nuclear equation for the decay of phosphorus-32. **1**

(*b*) (i) An 8 g sample of phosphorus-32 was freshly prepared.

Calculate the number of phosphorus atoms contained in the 8 g sample. **1**

(ii) The half-life of phosphorus-32 is 14·3 days.

Calculate the time it would take for the mass of phosphorus-32 in the 8 g sample to fall to 1 g. **1**

(3)

[Turn over

Marks

3. Sphalerite is an impure zinc sulphide ore, containing traces of other metal compounds.

The flow diagram for the extraction of zinc from this ore is shown below.

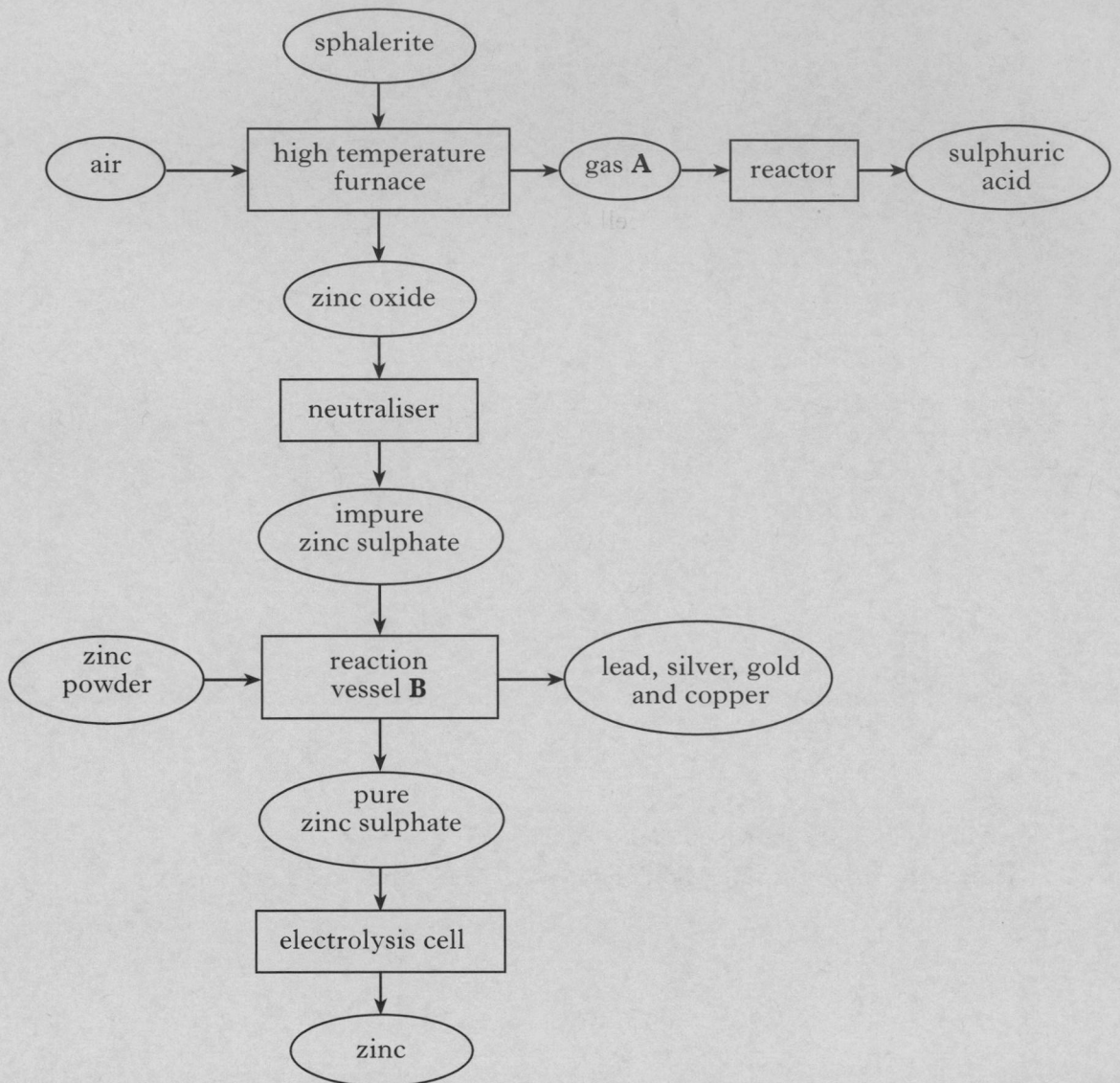

(*a*) Name gas **A**.

1

(*b*) Name the type of reaction taking place in reaction vessel **B**.

1

Page twelve

DO NOT WRITE IN THIS MARGIN

Marks

3. (continued)

(*c*) It is economical to make use of the sulphuric acid produced.

Add an arrow to the flow diagram to show how the sulphuric acid could be used in this extraction. **1**

(*d*) The ion-electron equation for the production of zinc in the electrolysis cell is

$$Zn^{2+} + 2e^{-} \longrightarrow Zn$$

If a current of 2000 A is used in the cell, calculate the mass of zinc, in kg, produced in 24 hours.

Show your working clearly. **2**

(5)

[Turn over

Marks

4. Glucose is produced in plants by photosynthesis.

(*a*) Plants convert glucose into a condensation polymer for storing energy. Name this condensation polymer.

1

(*b*) One way of representing the structure of glucose in aqueous solution is shown below.

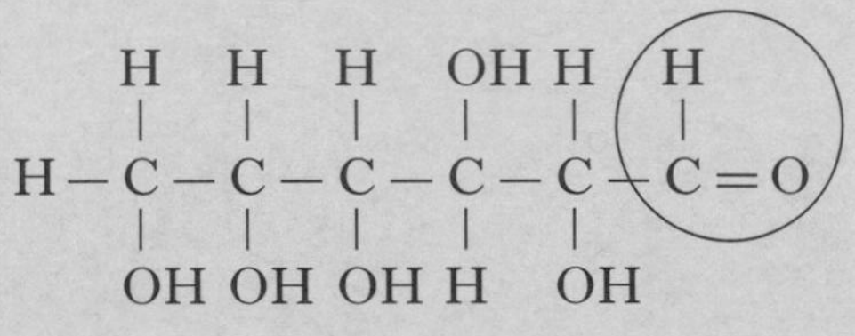

$$H-\overset{\overset{\displaystyle H}{|}}{C}-\overset{\overset{\displaystyle H}{|}}{C}-\overset{\overset{\displaystyle H}{|}}{C}-\overset{\overset{\displaystyle OH}{|}}{C}-\overset{\overset{\displaystyle H}{|}}{C}-\overset{\overset{\displaystyle H}{|}}{C}=O$$

In this structure the aldehyde group is circled.

(i) What would be seen when glucose is oxidised using Tollens' reagent?

1

(ii) Complete the structure below to show the product formed when glucose is oxidised.

$$H-\overset{\overset{\displaystyle H}{|}}{C}-\overset{\overset{\displaystyle H}{|}}{C}-\overset{\overset{\displaystyle H}{|}}{C}-\overset{\overset{\displaystyle OH}{|}}{C}-\overset{\overset{\displaystyle H}{|}}{C}-C$$

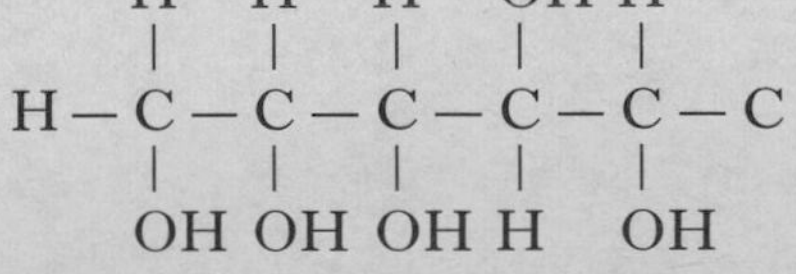

1

(*c*) Under anaerobic conditions, carbohydrates, like glucose, can be used to produce biogas. The main constituent of biogas is methane which is a useful fuel.

State **one** advantage of using biogas as a fuel rather than natural gas.

1

(4)

DO NOT WRITE IN THIS MARGIN

Marks

5. If the conditions are kept constant, reversible reactions will attain a state of equilibrium.

(*a*) Circle the correct words in the table to show what is true for reactions at equilibrium.

Rate of forward reaction compared to rate of reverse reaction	faster / same / slower
Concentrations of reactants compared to concentrations of products	usually different / always the same

1

(*b*) The following equilibrium involves two compounds of phosphorus.

$$PCl_3(g) + 3NH_3(g) \rightleftharpoons P(NH_2)_3(g) + 3HCl(g)$$

(i) An increase in temperature moves the above equilibrium to the left.

What does this indicate about the enthalpy change for the forward reaction?

1

(ii) What effect, if any, will an increase in pressure have on the above equilibrium?

1

(3)

[Turn over

Marks

6. The effect of temperature on reaction rate can be studied using the reaction between oxalic acid and acidified potassium permanganate solutions.

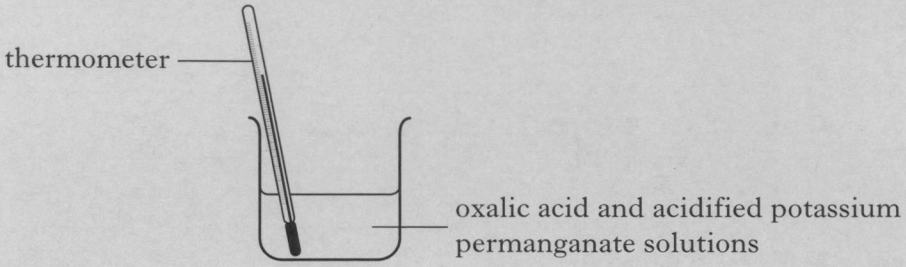

thermometer

oxalic acid and acidified potassium permanganate solutions

(a) What colour change would indicate that the reaction was complete?

1

(b) A student's results are shown on the graph below.

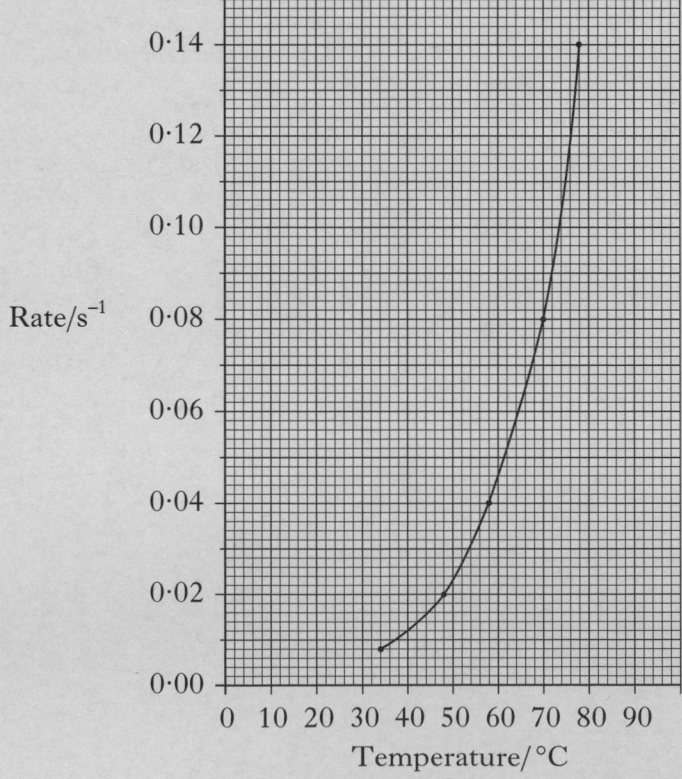

Rate/s⁻¹

Temperature/°C

(i) Use the graph to calculate the reaction time, in s, at 40 °C.

1

Marks

6. **(*b*) (continued)**

(ii) Why is it difficult to obtain an accurate reaction time when the reaction is carried out at room temperature?

1

(*c*) The diagram below shows the energy distribution of molecules in a gas at a particular temperature.

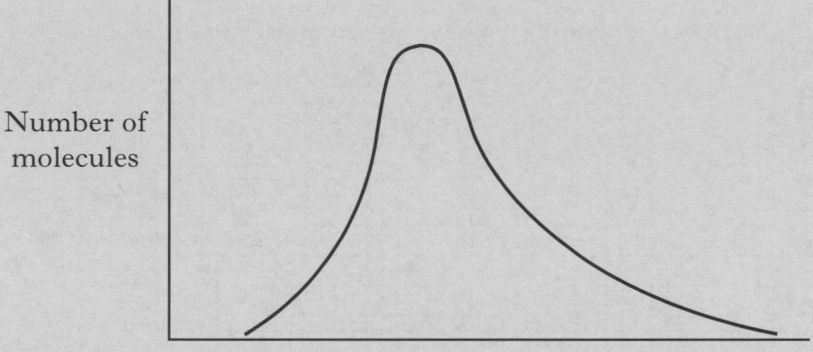

Number of molecules

Kinetic energy of the molecules

Draw a second curve on the diagram to show the energy distribution of the molecules in the gas at a higher temperature.

Label the diagram to indicate why an increase in temperature has such a significant effect on reaction rate.

1

(4)

[Turn over

Marks

7. Enzymes are specific biological catalysts. For example, trypsin, an enzyme produced in the pancreas, will catalyse the hydrolysis of only certain peptide links in a protein.

   (*a*) Draw the structure of a peptide link.

   1

   (*b*) Trypsin has an optimum temperature of 37 °C.

   Draw a curve to show how the enzyme activity varies with temperature.

   (Additional graph paper, if required, can be found on page 32.)

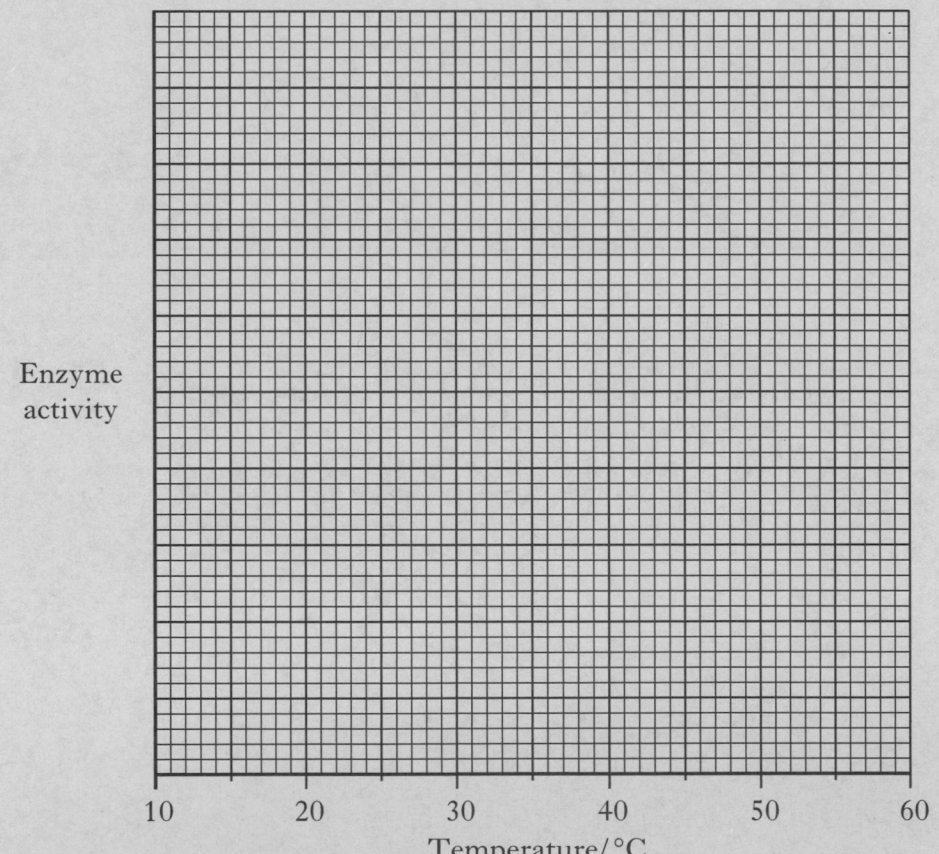

   Enzyme activity

   Temperature/°C

   1

   (*c*) Trypsin loses its activity if placed in a solution of very high pH.

   What happens to the enzyme to cause this loss of activity?

   1

   (3)

Marks

8. An experiment using dilute hydrochloric acid and sodium hydroxide solution was carried out to determine the enthalpy of neutralisation.

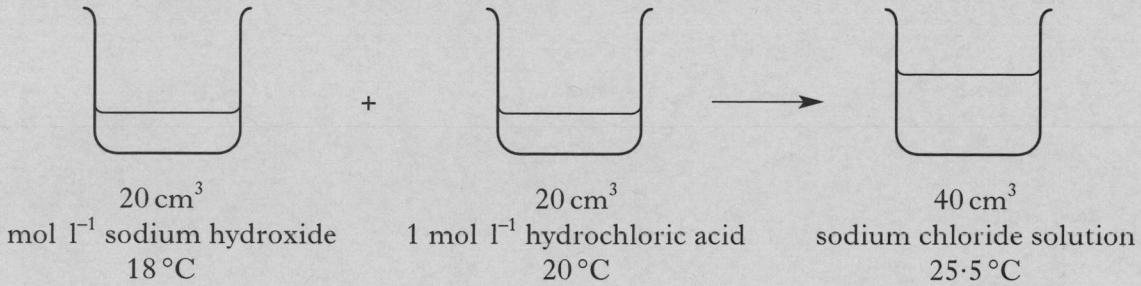

<table>
<tr><td>$20 \, cm^3$</td><td></td><td>$20 \, cm^3$</td><td>$40 \, cm^3$</td></tr>
<tr><td>$1 \, mol \, l^{-1}$ sodium hydroxide</td><td>+</td><td>$1 \, mol \, l^{-1}$ hydrochloric acid</td><td>sodium chloride solution</td></tr>
<tr><td>$18 \, ^{\circ}C$</td><td></td><td>$20 \, ^{\circ}C$</td><td>$25 \cdot 5 \, ^{\circ}C$</td></tr>
</table>

(*a*) Using the information in the diagram, calculate the enthalpy of neutralisation, in kJ mol^{-1} .

Show your working clearly.

2

(*b*) Calculate the concentration of hydroxide ions in the 1 mol l^{-1} hydrochloric acid used in the experiment.

1

(3)

[Turn over

Marks

9. (*a*) Aluminium and phosphorus are close to one another in the Periodic Table but the P^{3-} ion is much larger than the Al^{3+} ion.

Give the reason for this difference.

1

(*b*) The P^{3-} ion and the Ca^{2+} ion have the same electron arrangement but the Ca^{2+} ion is smaller than the P^{3-} ion.

Give the reason for this difference.

1

(2)

DO NOT WRITE IN THIS MARGIN

Marks

10. The structures of two antiseptics are shown. Both are aromatic.

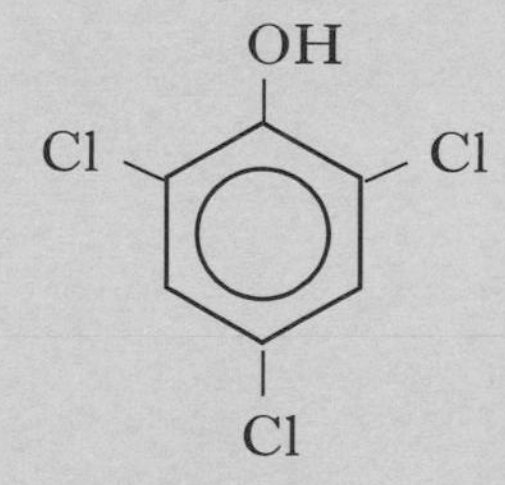

TCP

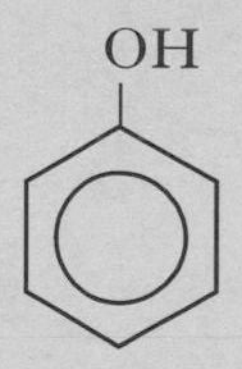

phenol

(*a*) (i) What gives the aromatic ring its stability? **1**

(ii) Write the molecular formula for TCP. **1**

(iii) The systematic name for TCP is 2,4,6-trichlorophenol.

The systematic name for Dettol, another antiseptic, is 4-chloro-3,5-dimethylphenol.

Draw a structural formula for Dettol. **1**

(*b*) The feedstocks for the production of antiseptics are made by reforming the naphtha fraction of crude oil.

Give another use for reformed naphtha. **1**

(4)

Marks

11. A student used the slow reaction between magnesium and water to determine the molar volume of hydrogen gas.

$$Mg(s) \quad + \quad 2H_2O(\ell) \quad \longrightarrow \quad Mg(OH)_2(aq) \quad + \quad H_2(g)$$

The following items were used in the experiment.

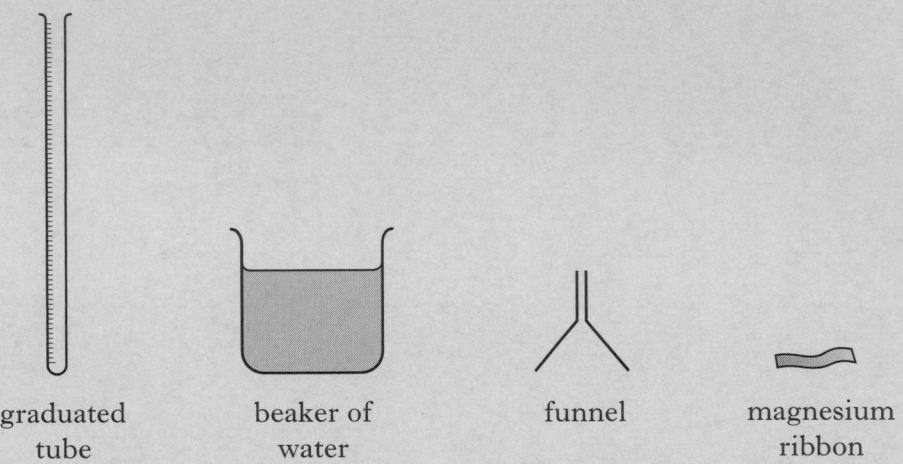

| graduated
tube | beaker of
water | funnel | magnesium
ribbon |

(*a*) Draw a diagram to show how the student would have arranged the above items at the start of the experiment.

1

DO NOT WRITE IN THIS MARGIN

Marks

11. (continued)

(*b*) What measurements would the student take and how would they be used to calculate the **molar** volume of hydrogen gas? **2**

(3)

[Turn over

Marks

12. A student added vitamin C solution to iodine solution.

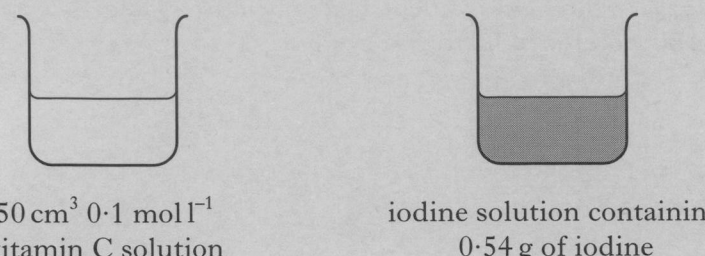

$50 \, cm^3 \, 0.1 \, mol \, l^{-1}$
vitamin C solution

iodine solution containing
$0.54 \, g$ of iodine

The equation for the reaction of vitamin C $(C_6H_8O_6)$ with iodine solution is shown below.

$$C_6H_8O_6(aq) \quad + \quad I_2(aq) \quad \longrightarrow \quad C_6H_6O_6(aq) \quad + \quad 2H^+(aq) \quad + \quad 2I^-(aq)$$
$$\text{(brown)} \qquad\qquad\qquad\qquad\qquad\qquad\qquad\qquad\qquad\qquad \text{(colourless)}$$

(a) By calculating which reactant was in excess, state whether the iodine solution would have been decolourised.

Show your working clearly.

2

(b) Write the ion-electron equation for the oxidation of vitamin C.

1

(3)

Marks

13. Compound **X** is a secondary alcohol.

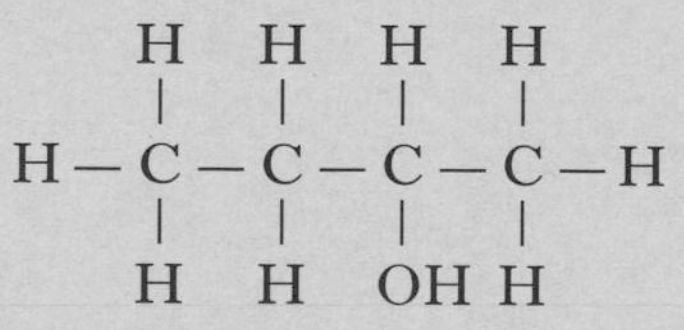

$$H-C-C-C-C-H$$

with H, H, H, H on top and H, H, OH, H on bottom

compound **X**

(a) Name compound **X**.

1

(b) Draw a structural formula for the tertiary alcohol that is an isomer of compound **X**.

1

(c) When passed over heated aluminium oxide, compound **X** is dehydrated, producing isomeric compounds, **Y** and **Z**.

Both compounds **Y** and **Z** react with hydrogen bromide, HBr. Compound **Y** reacts to produce two products while compound **Z** reacts to produce only one product.

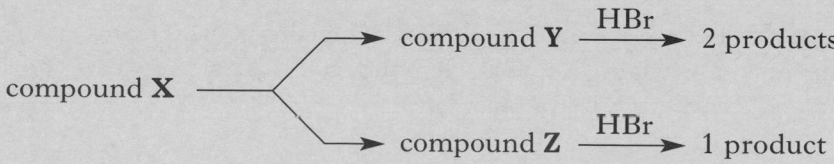

Name compound **Z**.

1

(3)

Marks

14. (*a*) Ethanol and propanoic acid can react to form an ester.

(i) Draw a structural formula for this ester.

1

(ii) Draw a labelled diagram of the assembled apparatus that could be used to prepare this ester in the laboratory.

2

(iii) Due to hydrogen bonding, ethanol and propanoic acid are soluble in water whereas the ester produced is insoluble.

In each of the boxes below, draw a molecule of water and use a dotted line to show where a hydrogen bond could exist between the organic molecule and the water molecule.

ethanol	propanoic acid
$CH_3 - CH_2 - O - H$	$CH_3 - CH_2 - \overset{\overset{\text{O}}{\|}}{C} - O - H$

1

DO NOT WRITE IN THIS MARGIN

Marks

14. (continued)

(*b*) Pyrolysis (thermal decomposition) of esters can produce two compounds, an alkene and an alkanoic acid, according to the following equation.

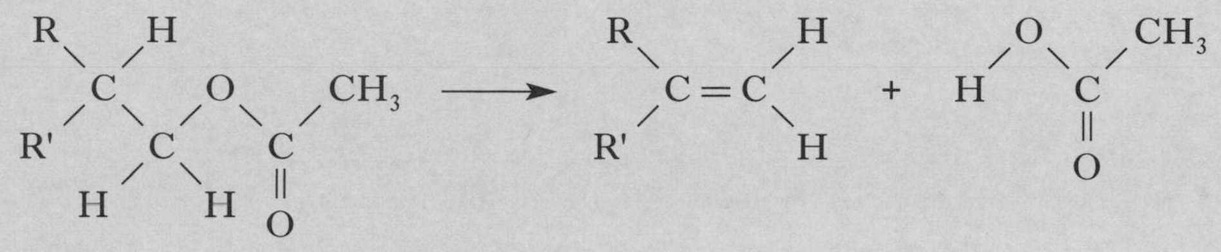

(R and R' represent alkyl groups)

Draw a structural formula for the ester that would produce 2-methylbut-1-ene and methanoic acid on pyrolysis. **1**

(5)

[Turn over

DO NOT WRITE IN THIS MARGIN

Marks

15. Vinegar is a dilute solution of ethanoic acid.

(*a*) Hess's Law can be used to calculate the enthalpy change for the formation of ethanoic acid from its elements.

$$\underset{\text{(graphite)}}{2C(s)} + 2H_2(g) + O_2(g) \rightarrow CH_3COOH(\ell)$$

Calculate the enthalpy change for the above reaction, in kJ mol^{-1}, using information from the data booklet and the following data.

$$CH_3COOH(\ell) + 2O_2(g) \rightarrow 2CO_2(g) + 2H_2O(\ell) \quad \Delta H = -876\,\text{kJ mol}^{-1}$$

Show your working clearly.

2

(*b*) Ethanoic acid can be used to prepare the salt, sodium ethanoate, CH_3COONa.

Explain why sodium ethanoate solution has a pH greater than 7.

In your answer you should mention the **two** equilibria involved.

3

(5)

DO NOT WRITE IN THIS MARGIN

Marks

16. Potassium permanganate is a very useful chemical in the laboratory.

(*a*) Solid potassium permanganate can be heated to release oxygen gas. This reaction can be represented by the equation shown below.

$$KMnO_4(s) \longrightarrow K_2O(s) + MnO_2(s) + O_2(g)$$

Balance the above equation. **1**

(*b*) An acidified potassium permanganate solution can be used to determine the concentration of a solution of iron(II) sulphate by a titration method.

(i) Apart from taking accurate measurements, suggest **two** points of good practice that a student should follow to ensure that an accurate end-point is achieved in a titration.

1

(ii) In a titration, a student found that an average of $16{\cdot}7\,cm^3$ of iron(II) sulphate solution was needed to react completely with $25{\cdot}0\,cm^3$ of $0{\cdot}20\,mol\,l^{-1}$ potassium permanganate solution.

The equation for the reaction is:

$$5Fe^{2+}(aq) + MnO_4^{-}(aq) + 8H^{+}(aq) \rightarrow 5Fe^{3+}(aq) + Mn^{2+}(aq) + 4H_2O(\ell)$$

Calculate the concentration of the iron(II) sulphate solution, in $mol\,l^{-1}$.

Show your working clearly.

2

(4)

Marks

17. A proton NMR spectrum can be used to help identify the structure of an organic compound.

The three key principles used in identifying a group containing hydrogen atoms in a molecule are as follows:

- The position of the line(s) on the x-axis of the spectrum is a measure of the "chemical shift" of the hydrogen atoms in the particular group.

Some common "chemical shift" values are given in the table below.

Group containing hydrogen atoms	Chemical shift
$-CH_3$	1·0
$-C \equiv CH$	2·7
$-CH_2Cl$	3·7
$-CHO$	9·0

- The number of lines for the hydrogen atoms in the group is n + 1 where n is the number of hydrogen atoms on the carbon atom next to the group.

- The maximum height of the line(s) for the hydrogen atoms in the group is relative to the number of hydrogen atoms in the group.

The spectrum for ethanal is shown below.

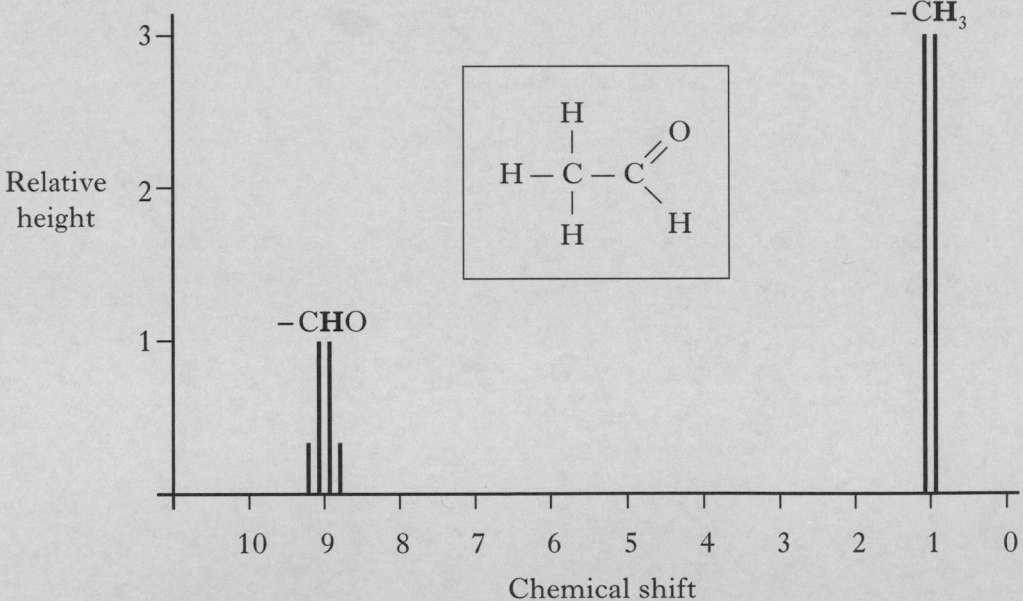

(a) The chemical shift values shown in the table are based on the range of values shown in the data booklet for proton NMR spectra.

Use the data booklet to find the range in the chemical shift values for hydrogen atoms in the following environment:

$$\begin{array}{c} \diagdown \\ C = C \\ \diagup \quad \diagdown_H \end{array}$$

1

Marks

17. (continued)

(*b*) A carbon compound has the following spectrum.

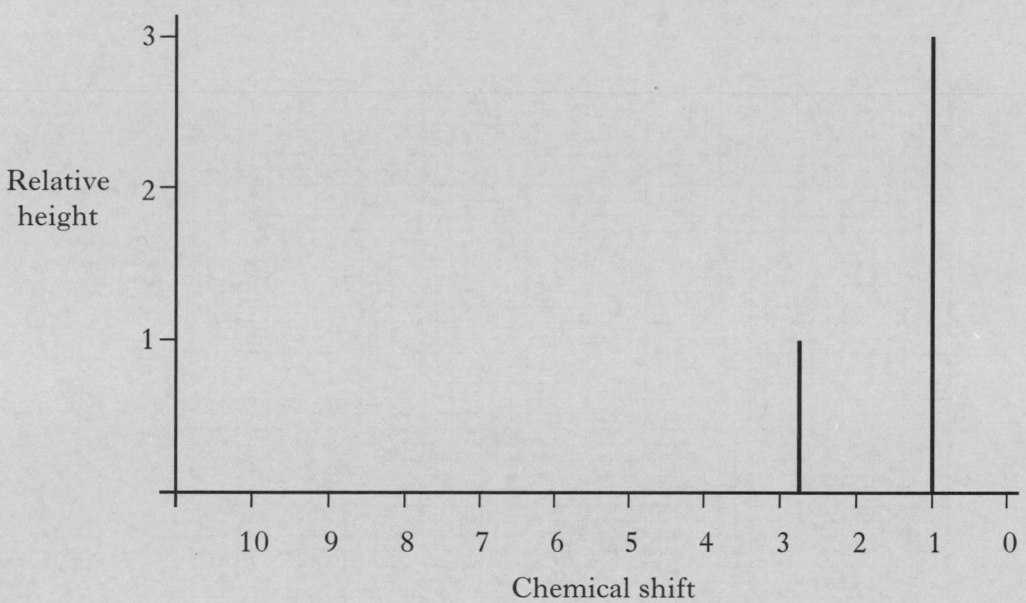

Name this compound.

1

(*c*) Draw the spectrum that would be obtained for chloroethane.

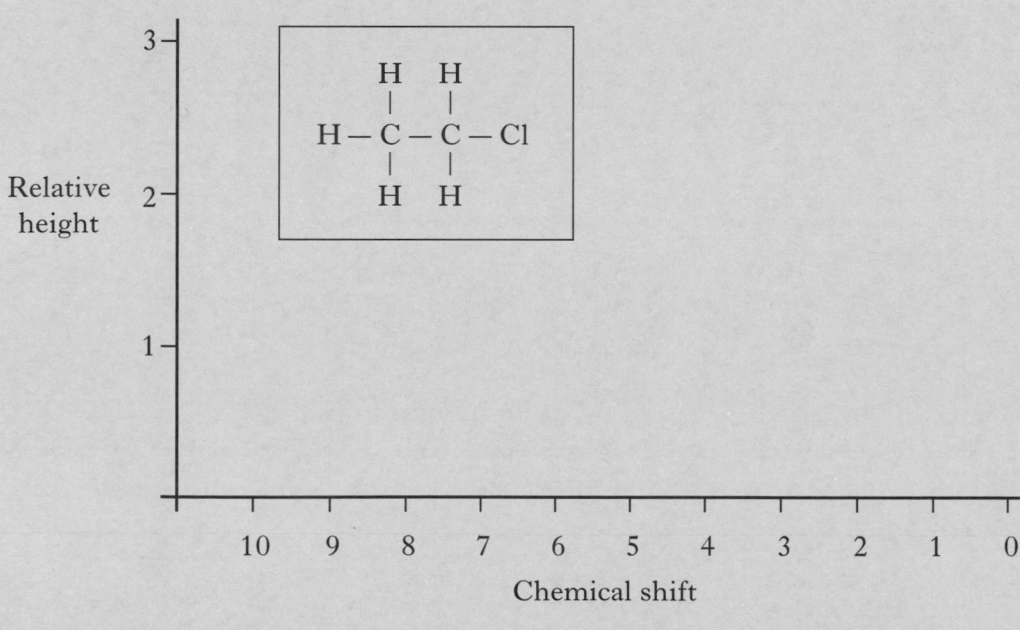

1

(3)

[*END OF QUESTION PAPER*]

ADDITIONAL SPACE FOR ANSWERS

ADDITIONAL GRAPH PAPER FOR QUESTION 7(*b*)

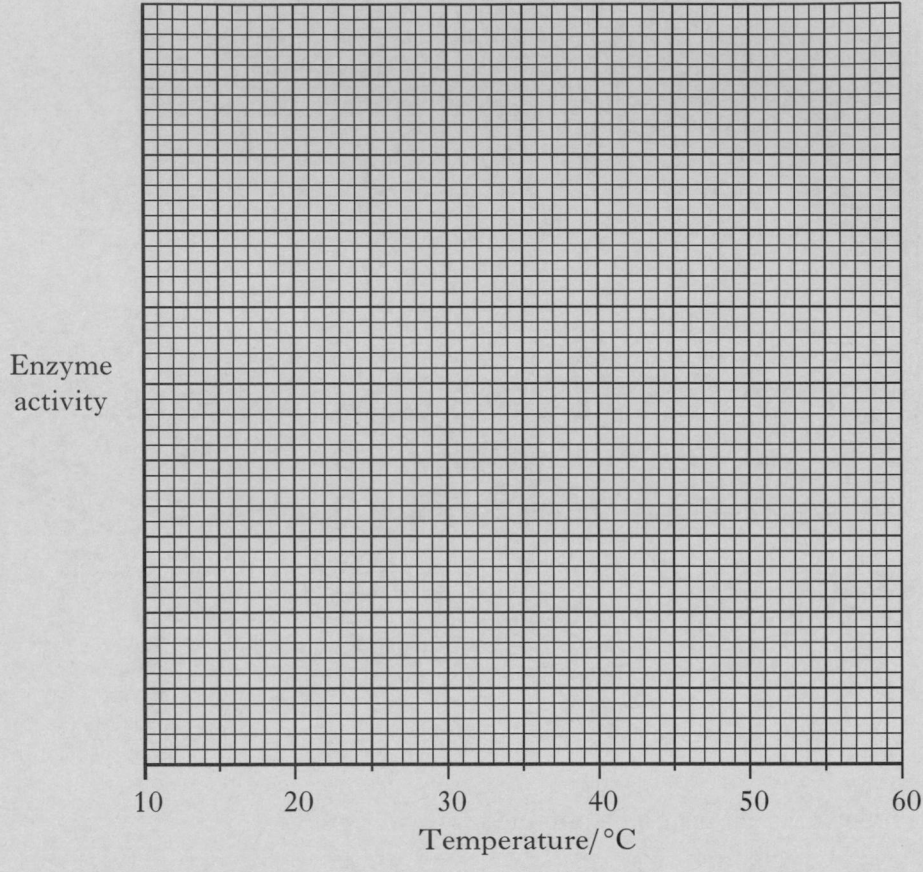

[BLANK PAGE]

FOR OFFICIAL USE

Total
Section B

X012/301

NATIONAL
QUALIFICATIONS
2005

TUESDAY, 31 MAY
9.00 AM – 11.30 AM

CHEMISTRY
HIGHER

Fill in these boxes and read what is printed below.

Full name of centre

Town

Forename(s)

Surname

Date of birth

Day Month Year   Scottish candidate number   Number of seat

Reference may be made to the Chemistry Higher and Advanced Higher Data Booklet (1999 edition).

SECTION A—Questions 1–40 (40 marks)

Instructions for completion of **Section A** are given on page two.

SECTION B (60 marks)

1   All questions should be attempted.

2   The questions may be answered in any order but all answers are to be written in the spaces provided in this answer book, and must be written clearly and legibly in ink.

3   Rough work, if any should be necessary, should be written in this book and then scored through when the fair copy has been written. If further space is required, a supplementary sheet for rough work may be obtained from the invigilator.

4   Additional space for answers will be found at the end of the book. If further space is required, supplementary sheets may be obtained from the invigilator and should be inserted inside the **front** cover of this book.

5   The size of the space provided for an answer should not be taken as an indication of how much to write. It is not necessary to use all the space.

6   Before leaving the examination room you must give this book to the invigilator. If you do not, you may lose all the marks for this paper.

SCOTTISH
QUALIFICATIONS
AUTHORITY

SECTION A

Read carefully

1 Check that the answer sheet provided is for **Chemistry Higher (Section A)**.

2 Check that the answer sheet you have been given has **your name**, **date of birth**, **SCN** (Scottish Candidate Number) and **Centre Name** printed on it.

Do not change any of these details.

3 If any of this information is wrong, tell the Invigilator immediately.

4 If this information is correct, **print** your name and seat number in the boxes provided.

5 Use **black** or **blue ink** for your answers. **Do not use red ink**.

6 The answer to each question is **either** A, B, C or D. Decide what your answer is, then put a horizontal line in the space provided (see sample question below).

7 There is **only one correct** answer to each question.

8 Any rough working should be done on the question paper or the rough working sheet, **not** on your answer sheet.

9 At the end of the exam, put the **answer sheet for Section A inside the front cover of your answer book**.

Sample Question

To show that the ink in a ball-pen consists of a mixture of dyes, the method of separation would be

        A fractional distillation

        B chromatography

        C fractional crystallisation

        D filtration.

The correct answer is **B**—chromatography. The answer **B** has been clearly marked with a horizontal line (see below).

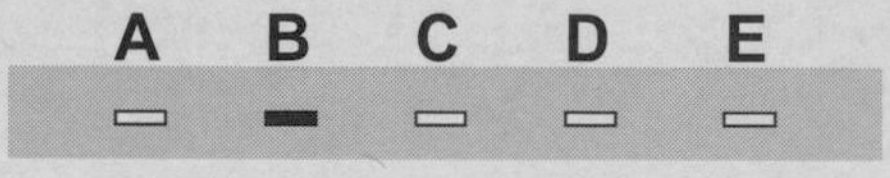

Changing an answer

If you decide to change your answer, cancel your first answer by putting a cross through it (see below) and fill in the answer you want. The answer below has been changed to **B**.

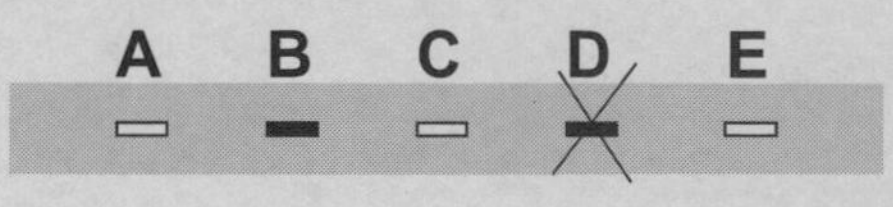

If you then decide to change back to an answer you have already scored out, put a tick (✓) to the **right** of the answer you want, as shown below:

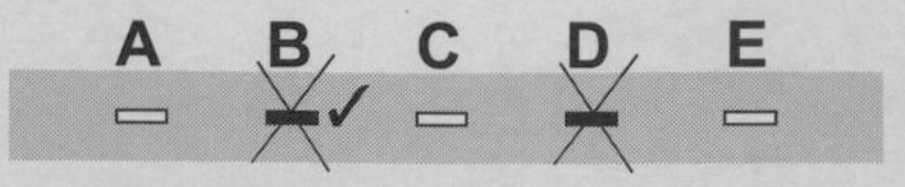

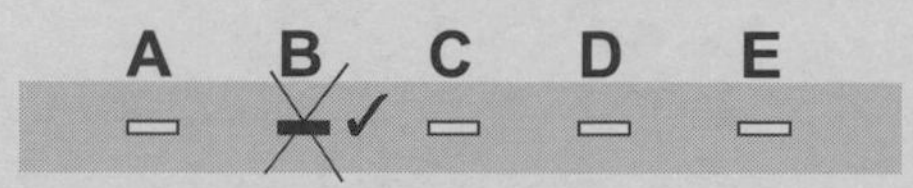

1. Isotopes of an element have

   A the same mass number

   B the same number of neutrons

   C equal numbers of protons and neutrons

   D different numbers of neutrons.

2. Which of the following pairs of solutions would react to produce a precipitate?

   A Barium nitrate and sodium chloride

   B Barium hydroxide and potassium nitrate

   C Copper(II) sulphate and sodium carbonate

   D Copper(II) chloride and potassium sulphate

3. Dilute hydrochloric acid, concentration $2 \, mol \, l^{-1}$, is added to a mixture of copper metal and copper(II) carbonate.

   Which of the following happens?

   A The only gas produced is carbon dioxide.

   B The only gas produced is hydrogen.

   C A mixture of carbon dioxide and hydrogen is produced.

   D There is no production of gas.

4. How many moles of magnesium will react with $20 \, cm^3$ of $2 \, mol \, l^{-1}$ hydrochloric acid?

   $$Mg(s) + 2HCl(aq) \rightarrow MgCl_2(aq) + H_2(g)$$

   A 0·01

   B 0·02

   C 0·04

   D 0·20

5. The continuous use of large extractor fans greatly reduces the possibility of an explosion in a flour mill. This is mainly because

   A a build up in the concentration of oxygen is prevented

   B local temperature rises are prevented by the movement of the air

   C particles of flour suspended in the air are removed

   D the slow accumulation of carbon monoxide is prevented.

6.

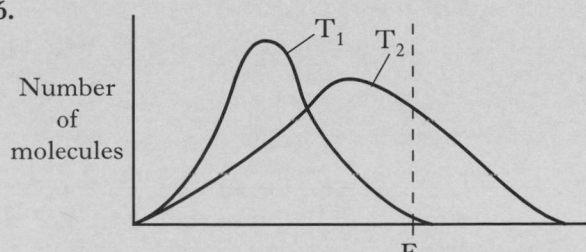

Kinetic energy of molecules

Which of the following is the correct interpretation of the above energy distribution diagram for a reaction as the temperature **decreases** from T_2 to T_1?

	Activation energy (E_A)	Number of successful collisions
A	remains the same	increases
B	decreases	decreases
C	decreases	increases
D	remains the same	decreases

7. Which of the following equations illustrates an enthalpy of combustion?

   A $C_2H_6 + 3\frac{1}{2}O_2(g)$
            $\downarrow$
            $2CO_2(g) + 3H_2O(\ell)$

   B $C_2H_5OH(\ell) + O_2(g)$
            $\downarrow$
            $CH_3COOH(\ell) + H_2O(\ell)$

   C $CH_3CHO(\ell) + \frac{1}{2}O_2(g)$
            $\downarrow$
            $CH_3COOH(\ell)$

   D $CH_4(g) + 1\frac{1}{2}O_2(g)$
            $\downarrow$
            $CO(g) + 2H_2O(\ell)$

[Turn over

8. The bond enthalpy of a gaseous diatomic molecule is the energy required to break one mole of the covalent bonds. It is also the energy released in the formation of one mole of the bonds from the atoms involved.

Bond	Bond enthalpy/kJ mol^{-1}
H — H	432
I — I	149
H — I	295

$$H_2(g) + I_2(g) \rightarrow 2HI(g)$$

What is the enthalpy change, in kJ mol^{-1}, for the above reaction?

A   +9

B   −9

C   +286

D   −286

9. Which of the following equations represents the first ionisation energy of chlorine?

A $Cl(g) + e^- \rightarrow Cl^-(g)$

B $Cl^+(g) + e^- \rightarrow Cl(g)$

C $Cl(g) \rightarrow Cl^+(g) + e^-$

D $Cl^-(g) \rightarrow Cl(g) + e^-$

10. Which of the following elements has the smallest electronegativity?

A   Lithium

B   Caesium

C   Fluorine

D   Iodine

11. A substance melts at 1074 °C and boils at 1740 °C. The passage of an electric current through the molten substance results in electrolysis.

What type of structure is present in the substance?

A   Ionic

B   Metallic

C   Covalent molecular

D   Covalent network

12. Which of the following occurs when crude oil is distilled?

A   Covalent bonds break and form again.

B   Van der Waals' bonds break and form again.

C   Covalent bonds break and van der Waals' bonds form.

D   Van der Waals' bonds break and covalent bonds form.

13. Which of the following has a covalent molecular structure?

A   Radium chloride

B   A noble gas

C   Silicon dioxide

D   A fullerene

14. A metal (melting point 328 °C, density 11·3 g cm^{-3}) was obtained by electrolysis of its molten chloride (melting point 501 °C, density 5·84 g cm^{-3}).

During the electrolysis, how would the metal occur?

A   As a solid on the surface of the electrolyte

B   As a liquid on the surface of the electrolyte

C   As a solid at the bottom of the electrolyte

D   As a liquid at the bottom of the electrolyte

15. Which of the following gases contains the smallest number of molecules?

A   100 g fluorine

B   100 g nitrogen

C   100 g oxygen

D   100 g hydrogen

16. Approximately how many atoms will be present in 11·5 litres of carbon monoxide?

(Take the molar volume of carbon monoxide to be 23 litres mol^{-1}.)

A $1·5 \times 10^{23}$

B 3×10^{23}

C 6×10^{23}

D $1·2 \times 10^{24}$

17. $3CuO + 2NH_3 \rightarrow 3Cu + N_2 + 3H_2O$

What volume of gas, in cm^3, would be obtained by reaction between $100\,cm^3$ of ammonia gas and excess copper(II) oxide?

(All volumes are measured at atmospheric pressure and $20\,°C$.)

A   50

B   100

C   200

D   400

18. The following reaction

$CH_3-CH_2-CH_2-CH_2-CH_2-CH_3 \longrightarrow$ 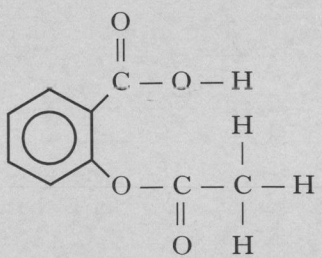 $+ 4H_2$

can take place during

A   dehydration

B   cracking

C   hydrogenation

D   reforming.

19. Biogas is produced under anaerobic conditions by the fermentation of biological materials.

What is the main constituent of biogas?

A   Butane

B   Ethane

C   Methane

D   Propane

20. Which of the following organic compounds is an isomer of hexanal?

A   2-Methylbutanal

B   3-Methylpentan-2-one

C   2,2-Dimethylbutan-1-ol

D   3-Ethylpentanal

21. Aspirin is one of the most widely used pain relievers in the world. It has the structure:

Which **two** functional groups are present in an aspirin molecule?

A   Aldehyde and ketone

B   Carboxyl and ester

C   Ester and aldehyde

D   Hydroxyl and carboxyl

22. Which of the following hydrocarbons always gives the same product when one of its hydrogen atoms is replaced by a chlorine atom?

A   Hexane

B   Hex-1-ene

C   Cyclohexane

D   Cyclohexene

23. Oxidation of 4-methylpentan-2-ol using copper(II) oxide results in the alcohol

A   losing $2\,g$ per mole

B   gaining $2\,g$ per mole

C   gaining $16\,g$ per mole

D   not changing in mass.

24. Ozone has an important role in the upper atmosphere because it

A   reflects certain CFCs

B   absorbs certain CFCs

C   reflects ultraviolet radiation

D   absorbs ultraviolet radiation.

25. Synthesis gas consists mainly of

A CH_4 alone

B CH_4 and CO

C   CO and H_2

D CH_4, CO and H_2.

26. The following monomers can be used to prepare nylon–6,6.

$$Cl - \overset{\overset{\displaystyle O}{\|}}{C} - (CH_2)_4 - \overset{\overset{\displaystyle O}{\|}}{C} - Cl \quad \text{and} \quad H_2N - (CH_2)_6 - NH_2$$

What molecule is released during the polymerisation reaction between these monomers?

A HCl

B H_2O

C NH_3

D HOCl

27. Which of the following statements about nylon and polystyrene is true?

A Both are thermosetting plastics.

B Both are condensation polymers.

C Both give off carbon dioxide and water vapour on burning.

D Both have hydrogen bonds between the polymer chains.

28. Proteins can be denatured under acid conditions.

During this denaturing, the protein molecule

A changes shape

B is dehydrated

C is neutralised

D is polymerised.

29.

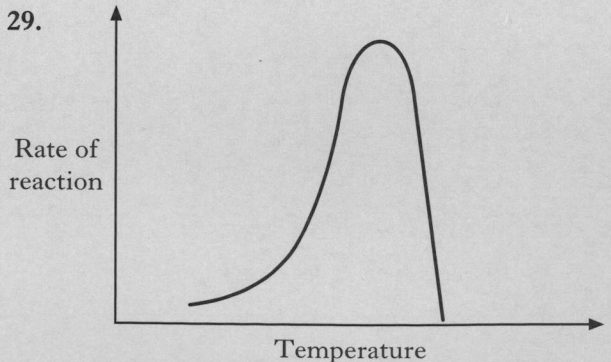

The above diagram could represent

A fermentation of glucose

B neutralisation of an acid by an alkali

C combustion of sucrose

D reaction of a metal with acid.

30. The costs involved in the industrial production of a chemical are made up of fixed costs and variable costs.

Which of the following is most likely to be classified as a variable cost?

A The cost of land rental

B The cost of plant construction

C The cost of labour

D The cost of raw materials

31. Which of the following is produced by a batch process?

A Sulphuric acid from sulphur and oxygen

B Aspirin from salicylic acid

C Iron from iron ore

D Ammonia from nitrogen and hydrogen

32. Consider the reaction pathway shown below.

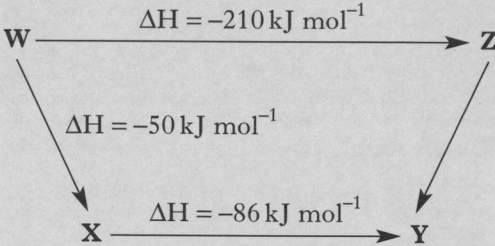

According to Hess's Law, the ΔH value, in $kJ\,mol^{-1}$, for reaction **Z** to **Y** is

A +74

B −74

C +346

D −346.

33. Chemical reactions are in a state of dynamic equilibrium only when

A the reaction involves zero enthalpy change

B the concentrations of reactants and products are equal

C the rate of the forward reaction equals that of the backward reaction

D the activation energies of the forward and backward reactions are equal.

34.

$$Cl_2(g) + H_2O(\ell) \rightleftharpoons Cl^-(aq) + ClO^-(aq) + 2H^+(aq)$$

The addition of which of the following substances would move the above equilibrium to the right?

A   Hydrogen

B   Hydrogen chloride

C   Sodium chloride

D   Sodium hydroxide

35. A trout fishery owner added limestone to his loch to combat the effects of acid rain. He managed to raise the pH of the water from 4 to 6.

The concentration of the $H^+(aq)$

A   increased by a factor of 2

B   increased by a factor of 100

C   decreased by a factor of 2

D   decreased by a factor of 100.

36. The concentration of $OH^-(aq)$ ions in a solution is $0.1\ mol\,l^{-1}$.

What is the pH of the solution?

A   1

B   8

C   13

D   14

37. A white solid dissolves in water, giving an alkaline solution, and reacts with dilute hydrochloric acid, giving off a gas.

(You may wish to refer to the data booklet.)

The solid could be

A   copper(II) ethanoate

B   potassium carbonate

C   ammonium chloride

D   lead(II) carbonate.

38. Which of the following is a redox reaction?

A $Zn \quad + 2HCl \rightarrow ZnCl_2 + H_2$

B $NaOH + HCl \rightarrow NaCl + H_2O$

C $NiO \quad + 2HCl \rightarrow NiCl_2 + H_2O$

D $CuCO_3 + 2HCl \rightarrow CuCl_2 + H_2O + CO_2$

39. Phosphorus-32 is made by neutron capture for use as a tracer in phosphate fertilisers.

From which of the following isotopes is phosphorus-32 made by neutron capture?

A $^{31}_{16}S$

B $^{31}_{15}P$

C $^{31}_{14}Si$

D $^{32}_{16}S$

40. The chart below was obtained from an 8-day old sample of an α-emitting radioisotope.

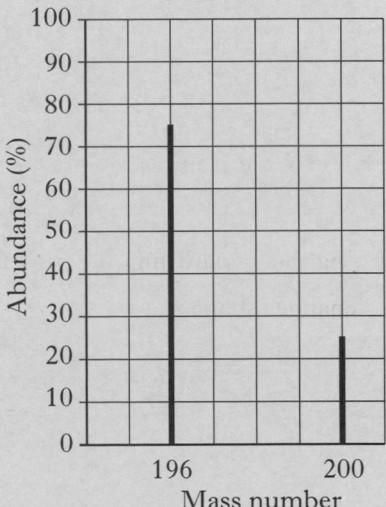

What is the half-life of the radioisotope?

A   2 days

B   4 days

C   8 days

D   12 days

Candidates are reminded that the answer sheet MUST be returned INSIDE the front cover of this answer book.

[Turn over for SECTION B on *Page eight*

Marks

SECTION B

1. The structure of a fat molecule is shown below.

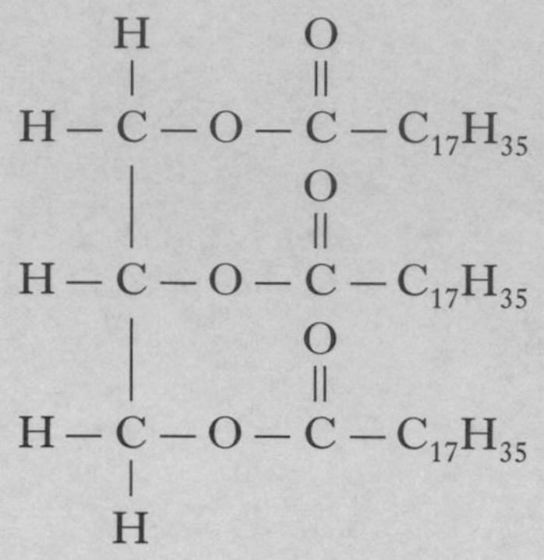

$$\begin{array}{c} H \\ | \\ H-C-O-C-C_{17}H_{35} \\ | \quad\quad\; \| \\ \quad\quad\; O \end{array}$$

H — C — O — C — C$_{17}$H$_{35}$
$\quad\quad\quad\quad\quad\;\; \|$
$\quad\quad\quad\quad\quad\;\; O$

H — C — O — C — C$_{17}$H$_{35}$
$\quad\quad\quad\quad\quad\;\; \|$
$\quad\quad\quad\quad\quad\;\; O$

H — C — O — C — C$_{17}$H$_{35}$
$\;\;|$
$\;\;H$

(a) When the fat is hydrolysed, a fatty acid is obtained.

Name the other product obtained in this reaction.

1

(b) Oils are liquid at room temperature; fats are solid.

Why do oils have lower melting points than fats?

1

(2)

Marks

2. The diagram below shows energy changes **A**, **B** and **C** for a reversible reaction.

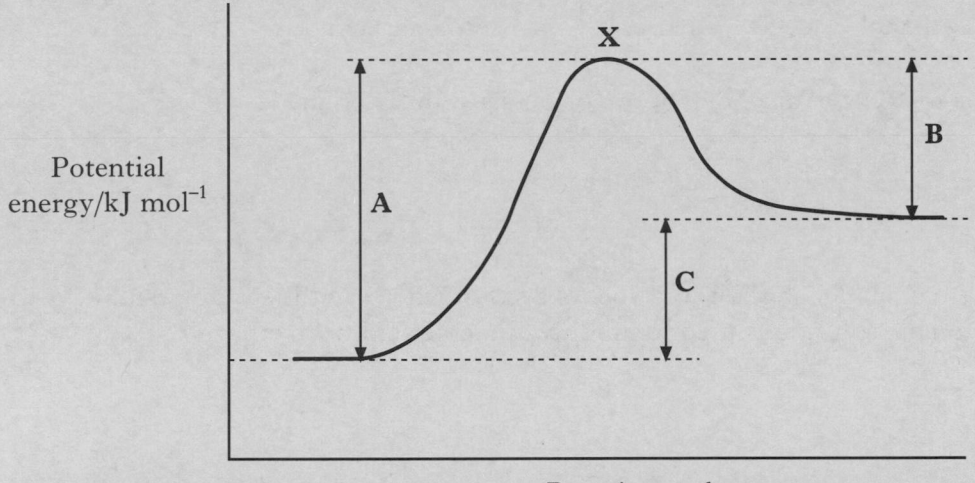

Potential
energy/kJ mol^{-1}

Reaction pathway

(a) What could be used to decrease both **A** and **B** but **not** change **C**?

1

(b) Give the name of the unstable arrangement of atoms formed at point **X**.

1

(2)

[Turn over

Marks

3. (*a*) Acidified potassium dichromate solution can be used to oxidise some alcohols to aldehydes and then to carboxylic acids, eg

$$\text{ethanol} \longrightarrow \text{ethanal} \longrightarrow \text{ethanoic acid}$$

(i) Name the type of alcohol that can be oxidised to an aldehyde.

1

(ii) What colour change would be observed when acidified potassium dichromate solution is used to produce ethanoic acid from ethanal?

1

(*b*) Ethanol and ethanoic acid react to form the ester, ethyl ethanoate.

$$\text{ethanol} \quad + \quad \text{ethanoic acid} \rightleftharpoons \text{ethyl ethanoate} \quad + \quad \text{water}$$

ethanol	ethanoic acid	ethyl ethanoate
Mass of one mole	Mass of one mole	Mass of one mole
= 46 g	= 60 g	= 88 g

(i) Clearly describe how the reaction mixture would be heated in the laboratory formation of the ester.

1

(ii) Use the above information to calculate the percentage yield of ethyl ethanoate if 5·0 g of ethanol produced 5·8 g of ethyl ethanoate on reaction with excess ethanoic acid.

Show your working clearly.

2

(5)

DO NOT WRITE IN THIS MARGIN

Marks

4. Urea, H_2NCONH_2, has several uses, eg as a fertiliser and for the manufacture of some thermosetting plastics.

It is produced in a two-step process.

Step one

$$CO_2(g) + 2NH_3(\ell) \rightleftharpoons \underset{\text{ammonium carbamate}}{H_2NCOONH_4(s)} \quad \Delta H \text{ negative}$$

Step two

$$\underset{\text{ammonium carbamate}}{H_2NCOONH_4(s)} \rightleftharpoons \underset{\text{urea}}{H_2NCONH_2(s)} + H_2O(g) \quad \Delta H \text{ positive}$$

(*a*) (i) What would happen to the equilibrium position in **step one** if the temperature was increased? **1**

(ii) **Step two** is carried out at low pressure.

Why does lowering the pressure move the equilibrium position to the right? **1**

(*b*) Draw the full structure for the negative ion in ammonium carbamate. **1**

(3)

[Turn over

Marks

5. The rate of carbon dioxide production was measured in three laboratory experiments carried out at the same temperature and using excess calcium carbonate.

Experiment	Acid	Calcium carbonate
A	40 cm³ of 0·10 mol l⁻¹ sulphuric acid	1 g lumps
B	40 cm³ of 0·10 mol l⁻¹ sulphuric acid	1 g powder
C	40 cm³ of 0·10 mol l⁻¹ hydrochloric acid	1 g lumps

The curve obtained for Experiment **A** is shown.

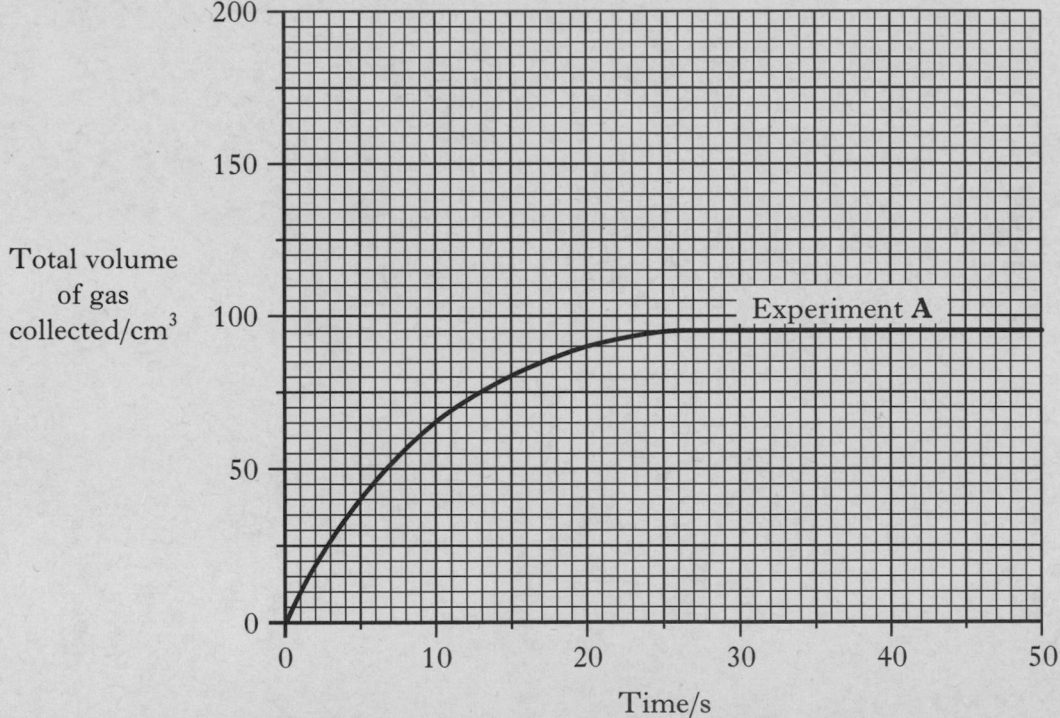

(a) Use the graph to calculate the average reaction rate, in cm³ s⁻¹, between 10 and 20 s.

1

DO NOT WRITE IN THIS MARGIN

Marks

5. (continued)

(*b*) Draw curves on the graph to show results that could be obtained for experiments **B** and **C**.

Label each curve clearly. 2

(Additional graph paper, if required, can be found on page 31.)

(*c*) Draw a labelled diagram of the assembled apparatus which could be used to carry out this experiment.

2

(5)

[Turn over

Marks

6. Uranium ore is converted into uranium(IV) fluoride, UF_4, to produce fuel for nuclear power stations.

(*a*) In one process, uranium can be extracted from the uranium(IV) fluoride by a redox reaction with magnesium, as follows.

$$2Mg \ + \ UF_4 \ \rightarrow \ 2MgF_2 \ + \ U$$

(i) Give another name for this type of redox reaction.

1

(ii) Write the ion-electron equation for the reduction reaction that takes place.

1

(iii) The reaction with magnesium is carried out at a high temperature.

The reaction vessel is filled with argon rather than air.

Suggest a reason for using argon rather than air.

1

Marks

6. (continued)

(b) In a second process, the uranium(IV) fluoride is converted into UF_6 as shown.

$$UF_4(s) \quad + \quad F_2(g) \quad \rightarrow \quad UF_6(g)$$

(i) Name the type of bonding in $UF_6(g)$.

1

(ii) Both UF_4 and UF_6 are radioactive.

How does the half-life of the uranium in UF_4 compare with the half-life of the uranium in UF_6?

1

(5)

[Turn over

Marks

7. The following apparatus can be used to determine the enthalpy of solution of a substance.

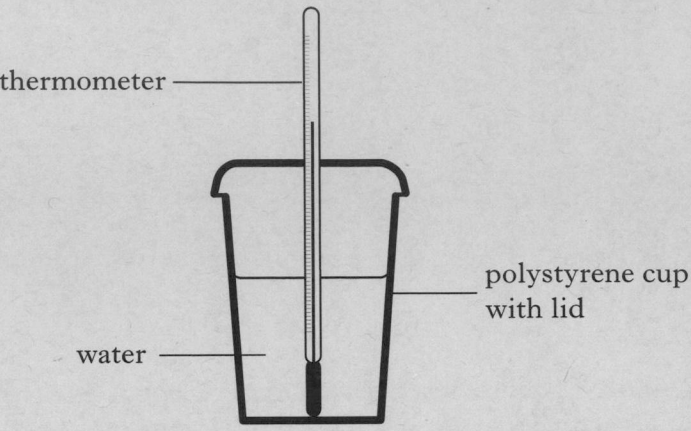

thermometer

polystyrene cup with lid

water

(a) Why was the experiment carried out in a polystyrene cup with a lid?

1

(b) In an experiment to find the enthalpy of solution of potassium hydroxide, KOH, a student added 3·6 g of the solid to the water in the polystyrene cup and measured the temperature rise. From this, it was calculated that the heat energy produced in the reaction was 3·5 kJ.

Use this information to calculate the enthalpy of solution of potassium hydroxide.

Show your working clearly.

2

(3)

Marks

8. Acid-base reactions are common in chemistry.

(a) Write the balanced equation for the reaction between copper(II) oxide and nitric acid.

1

(b) An acid can be thought of as a chemical which can release H^+ ions.

In an acid-base reaction the H^+ ions released by the acid are accepted by the base.

Some acid-base reactions are reversible. In these reactions both forward and reverse reactions involve the transfer of H^+ ions from the acid to the base.

(i) Using the information given above, complete the table showing the acid and base produced when HS^- ions react with H_3O^+ ions.

Acid	Base		Acid	Base
H_2O	NH_3	$\rightleftharpoons$	NH_4^+	OH^-
H_3O^+	HS^-	$\rightleftharpoons$		

1

(ii) Another reversible acid-base reaction is shown.

$$HCO_3^- \ + \ OH^- \ \rightleftharpoons \ CO_3^{2-} \ + \ H_2O$$

In the reverse reaction, state whether the water is acting as an acid or a base.

1

(3)

[Turn over

Marks

9. Hydrogen peroxide has a high viscosity.

The structure of hydrogen peroxide is shown below.

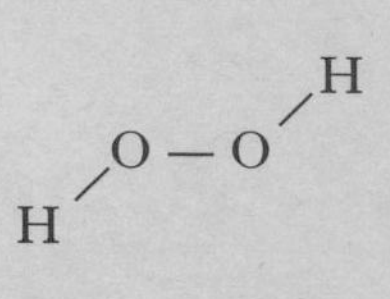

$$O - O$$ with H attached to each O

(*a*) Name the type of intermolecular force that is responsible for hydrogen peroxide's high viscosity.

1

(*b*) Hydrogen peroxide may be prepared from its elements.

The equation for the reaction is:

$$H_2(g) \ + \ O_2(g) \ \longrightarrow \ H_2O_2(\ell)$$

Calculate the enthalpy change, in kJ mol^{-1}, for the above reaction using the enthalpy of combustion of hydrogen from the data booklet and the enthalpy change for the following reaction.

$$H_2O_2(\ell) \ \longrightarrow \ \tfrac{1}{2}O_2(g) \ + \ H_2O(\ell) \qquad \Delta H = -98\,\text{kJ mol}^{-1}$$

Show your working clearly.

2

Marks

9. **(continued)**

(c) The following solution mixtures were used in a series of experiments involving the reaction between hydrogen peroxide in acid solution and potassium iodide solution.

$$H_2O_2(aq) \ + \ 2H^+(aq) \ + \ 2I^-(aq) \ \longrightarrow \ 2H_2O(\ell) \ + \ I_2(aq)$$

	Volume of KI(aq)/cm^3	Volume of H$_2$O(ℓ)/cm^3	Volume of H$_2$O$_2$(aq)/cm^3	Volume of H$_2$SO$_4$(aq)/cm^3	Volume of Na$_2$S$_2$O$_3$(aq)/cm^3	Rate/ s^{-1}
A	25	0	5	10	10	0·020
B	20	5	5	10	10	0·016
C	15	10	5	10	10	0·012
D	10	15	5	10	10	0·008
E	5	20	5	10	10	0·004

(i) From the information in the shaded columns in the table above, what variable is being kept constant throughout the series of experiments?

1

(ii) What was the aim of the series of experiments?

1

(iii) Calculate the time, in seconds, for the reaction in Experiment **A**.

1

(6)

Marks

10. Poly(ethenol) has an unusual property for a plastic.

    (a) What is this unusual property?

1

    (b) (i) A step in the manufacture of poly(ethenol) is shown below.

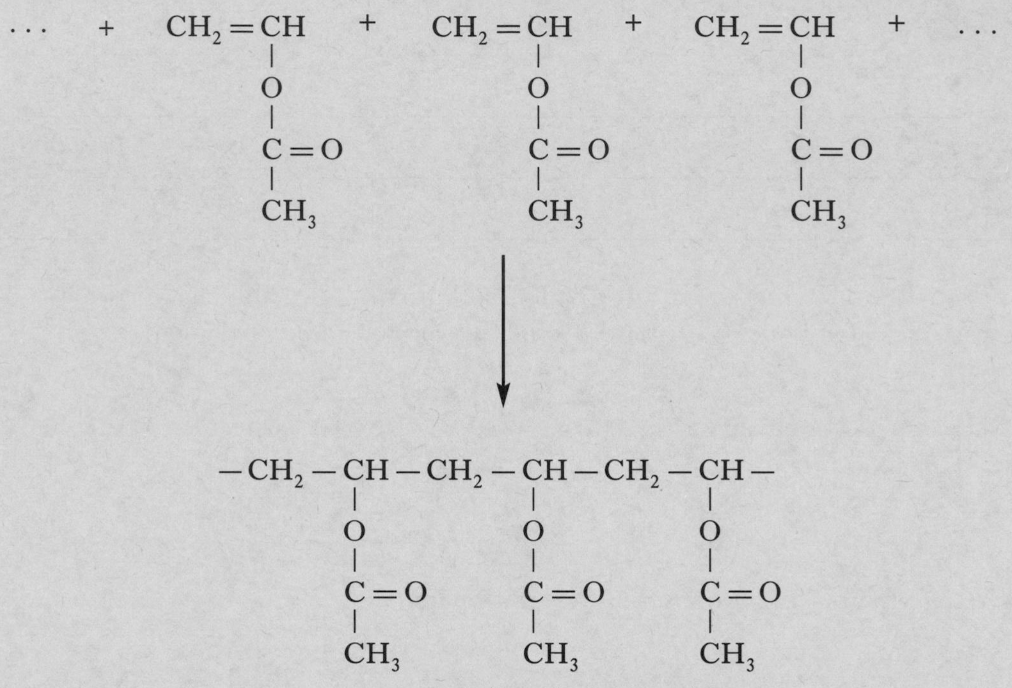

$$\cdots \ + \ CH_2\!=\!CH \ + \ CH_2\!=\!CH \ + \ CH_2\!=\!CH \ + \ \cdots$$

$$\begin{array}{ccc}
| & | & | \\
O & O & O \\
| & | & | \\
C\!=\!O & C\!=\!O & C\!=\!O \\
| & | & | \\
CH_3 & CH_3 & CH_3
\end{array}$$

$$-CH_2\!-\!CH\!-\!CH_2\!-\!CH\!-\!CH_2\!-\!CH\!-$$

$$\begin{array}{ccc}
| & | & | \\
O & O & O \\
| & | & | \\
C\!=\!O & C\!=\!O & C\!=\!O \\
| & | & | \\
CH_3 & CH_3 & CH_3
\end{array}$$

poly(ethenyl ethanoate)

Name the type of polymerisation which takes place in this step.

1

Marks

10. (continued)

(ii) The next step is ester exchange. This involves the removal of the ester side chains by reaction with an alcohol and sodium hydroxide.

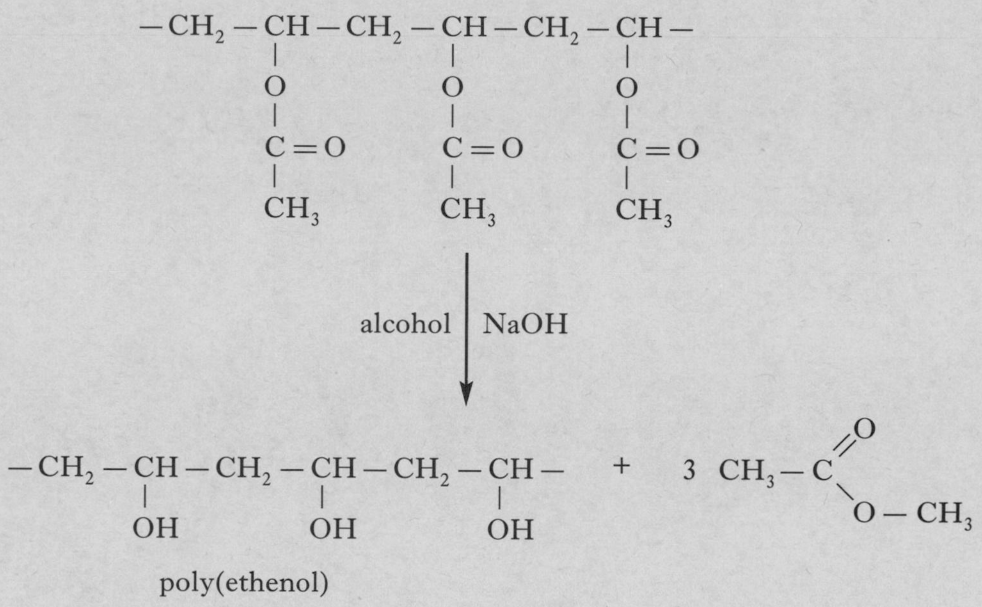

$$-CH_2-CH-CH_2-CH-CH_2-CH-$$

$$
\begin{array}{ccc}
O & O & O \\
| & | & | \\
C=O & C=O & C=O \\
| & | & | \\
CH_3 & CH_3 & CH_3
\end{array}
$$

alcohol | NaOH

$$-CH_2-CH-CH_2-CH-CH_2-CH- \quad + \quad 3\ CH_3-C\overset{O}{\underset{O-CH_3}{}}$$

$$
\begin{array}{ccc}
| & | & | \\
OH & OH & OH
\end{array}
$$

poly(ethenol)

Name the alcohol used in this step.

1

(3)

[Turn over

Marks

11. Respiration provides energy for the body through "combustion" of glucose.

The equation for the enthalpy of combustion of glucose is:

$$C_6H_{12}O_6(s) \; + \; 6O_2(g) \longrightarrow 6CO_2(g) \; + \; 6H_2O(\ell) \quad \Delta H = -2807\,kJ\,mol^{-1}$$

(*a*) Calculate the volume of oxygen, in litres, required to provide 418 kJ of energy.
(Take the molar volume of oxygen to be 24 litres mol^{-1}.)

Show your working clearly.

2

(*b*) In a poisonous atmosphere, a gas mask can be used to provide the oxygen needed for respiration. One type of gas mask contains potassium superoxide, KO_2, which reacts with water vapour to produce oxygen.

The balanced equation for the reaction is:

$$4KO_2(s) \; + \; 2H_2O(g) \; \rightarrow \; 4KOH(s) \; + \; 3O_2(g)$$

(i) Suggest why this reaction allows the same air to be breathed again and again.

1

(ii) Why is this type of mask also able to remove the carbon dioxide produced by respiration?

1

(4)

Marks

12. Ethanoic acid, $CH_3 - C \overset{O}{\underset{OH}{\diagup}}$ (aq), is a weak acid; hydrochloric acid, $HCl(aq)$, is a strong acid.

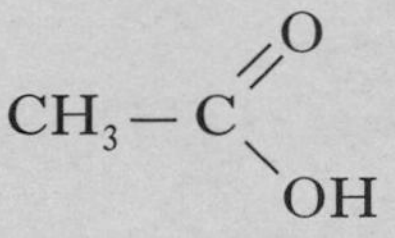

Using ethanoic acid and hydrochloric acid as examples, explain the differences in both pH and conductivity between $0.1 \, mol \, l^{-1}$ solutions of a strong and weak acid.

You may wish to use suitable equations in your answer.

(3)

[Turn over

DO NOT WRITE IN THIS MARGIN

Marks

13. Fuel cells can be used to power cars.

(*a*) (i) The ion-electron equations for the oxidation and reduction reactions that take place in a methanol fuel cell are:

$$CH_3OH(\ell) \quad + \quad H_2O(\ell) \quad \rightarrow \quad CO_2(g) \quad + \quad 6H^+(aq) \quad + \quad 6e^-$$

$$3O_2(g) \quad + \quad 12H^+(aq) \quad + \quad 12e^- \quad \rightarrow \quad 6H_2O(\ell)$$

Combine the two ion-electron equations to give the equation for the overall redox reaction. **1**

(ii) The equation for the overall redox reaction in a hydrogen fuel cell is

$$2H_2(g) \quad + \quad O_2(g) \quad \rightarrow \quad 2H_2O(\ell)$$

Give a disadvantage of the methanol fuel cell reaction compared to the hydrogen fuel cell reaction. **1**

DO NOT WRITE IN THIS MARGIN

Marks

13. (continued)

(*b*) The hydrogen gas for use in fuel cells can be produced by the electrolysis of water.

Hydrogen is produced at the negative electrode as shown.

$$2H_2O(\ell) \quad + \quad 2e^- \quad \rightarrow \quad H_2(g) \quad + \quad 2OH^-(aq)$$

Calculate the volume of hydrogen gas produced when a steady current of 0·50 A is passed through water for 30 minutes.

(Take the molar volume of hydrogen to be 24 litres mol^{-1}.)

Show your working clearly.

3

(5)

[Turn over

Marks

14. The structure of an ionic compound consists of a giant lattice of oppositely charged ions. The arrangement of ions is determined mainly by the "radius ratio" of the ions involved.

$$\text{radius ratio} = \frac{\text{radius of positive ion}}{\text{radius of negative ion}}$$

The arrangements for caesium chloride, CsCl, and sodium chloride, NaCl, are shown below.

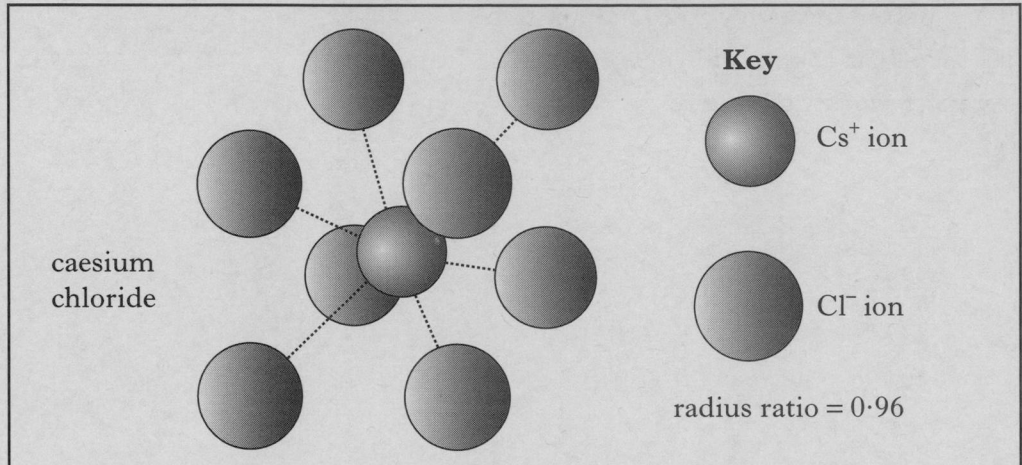

Key

Cs$^+$ ion

Cl$^-$ ion

caesium chloride

radius ratio = 0·96

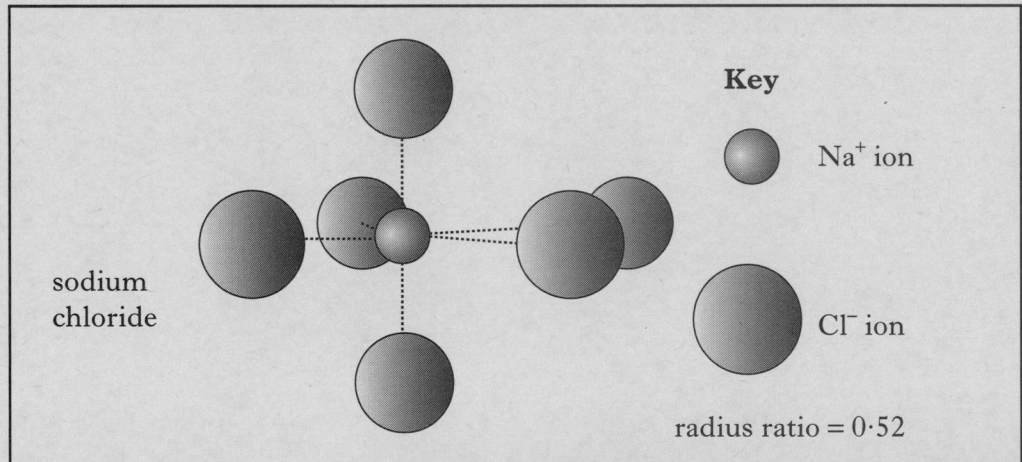

Key

Na$^+$ ion

Cl$^-$ ion

sodium chloride

radius ratio = 0·52

(*a*) By using the table of ionic radii on page 16 of the data booklet, calculate the radius ratio for magnesium oxide, MgO, and state which of the two arrangements, caesium chloride or sodium chloride, it is more likely to adopt.

1

DO NOT WRITE IN THIS MARGIN

Marks

14. (continued)

(*b*) The enthalpy of lattice breaking is the energy required to completely separate the ions from one mole of an ionic solid.

The table shows the enthalpies of lattice breaking, in kJ mol^{-1}, for some alkali metal halides.

Ions	F^-	Cl^-	Br^-
Li^+	1030	834	788
Na^+	910	769	732
K^+	808	701	671

Write a general statement linking the enthalpy of lattice breaking to ion size. **1**

(2)

[Turn over

Marks

15. Vitamin C is required by our bodies for producing the protein, collagen. Collagen can form sheets that support skin and internal organs.

(*a*) (i) There are two main types of protein.

Which of the two main types is collagen?

1

(ii) Part of the structure of collagen is shown.

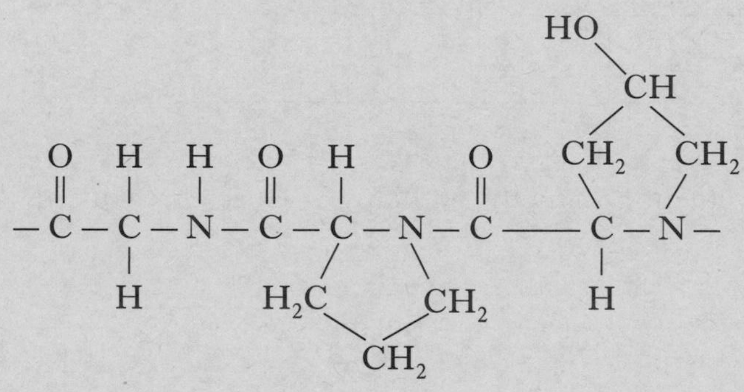

Draw a structural formula for an amino acid that could be obtained by hydrolysing this part of the collagen.

1

Marks

15. (continued)

(b) A standard solution of iodine can be used to determine the mass of vitamin C in orange juice.

Iodine reacts with vitamin C as shown by the following equation.

$$C_6H_8O_6(aq) \quad + \quad I_2(aq) \quad \rightarrow \quad C_6H_6O_6(aq) \quad + \quad 2H^+(aq) \quad + \quad 2I^-(aq)$$
vitamin C

In an investigation using a carton containing $500 \, cm^3$ of orange juice, separate $50.0 \, cm^3$ samples were measured out. Each sample was then titrated with a $0.0050 \, mol \, l^{-1}$ solution of iodine.

(i) Why would starch solution be added to each $50.0 \, cm^3$ sample of orange juice before titrating against iodine solution?

1

(ii) Titrating the whole carton of orange juice would require large volumes of iodine solution.

Apart from this disadvantage give another reason for titrating several smaller samples of orange juice.

1

(iii) An average of $21.4 \, cm^3$ of the iodine solution was required for the complete reaction with the vitamin C in $50.0 \, cm^3$ of orange juice.

Use this result to calculate the mass of vitamin C, in grams, in the $500 \, cm^3$ carton of orange juice.

Show your working clearly.

2

(6)

Marks

16. Carbon compounds take part in many different types of reactions.

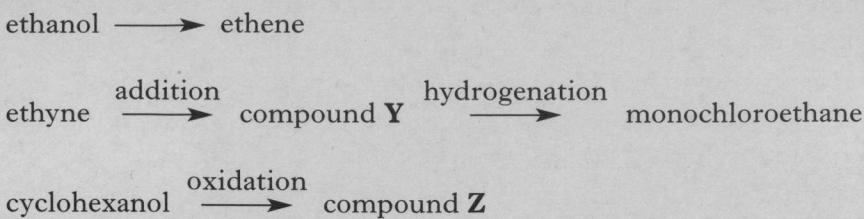

ethanol ⟶ ethene

ethyne $\xrightarrow{\text{addition}}$ compound **Y** $\xrightarrow{\text{hydrogenation}}$ monochloroethane

cyclohexanol $\xrightarrow{\text{oxidation}}$ compound **Z**

(a) Name the type of reaction that takes place in the formation of ethene from ethanol.

1

(b) Draw a structural formula for

   (i) compound **Y**;

1

   (ii) compound **Z**.

1

(3)

[END OF QUESTION PAPER]

ADDITIONAL SPACE FOR ANSWERS

ADDITIONAL GRAPH PAPER FOR QUESTION 5(*b*)

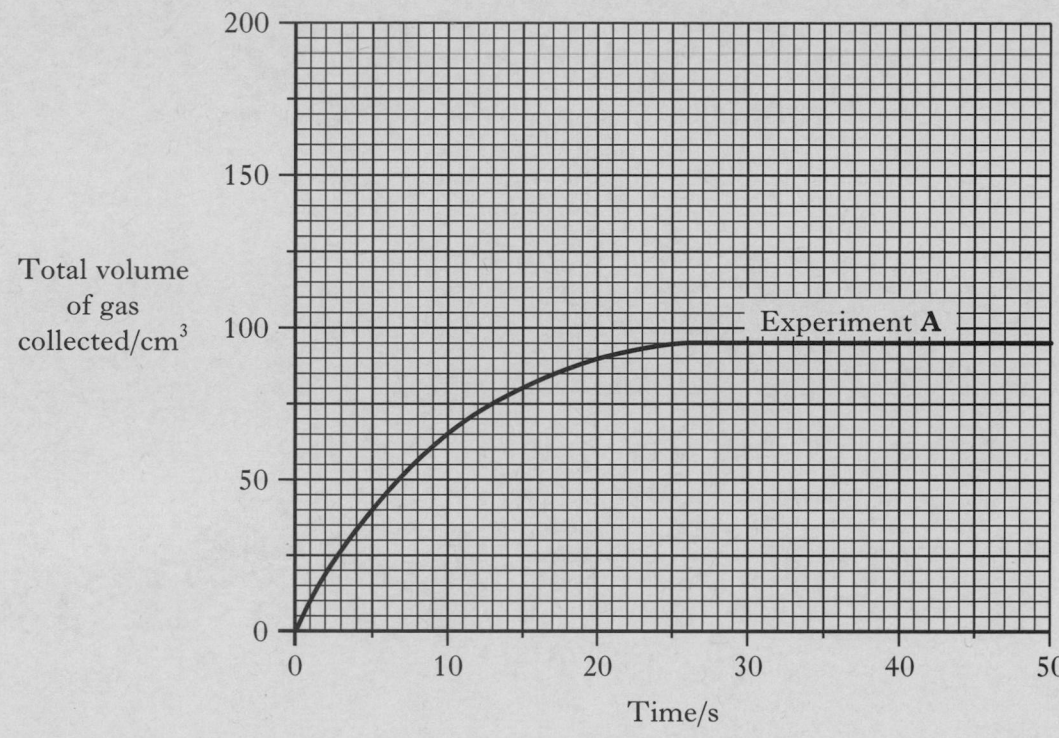

[BLANK PAGE]

[BLANK PAGE]

FOR OFFICIAL USE

Total
Section B

X012/301

NATIONAL
QUALIFICATIONS
2006

TUESDAY, 30 MAY
9.00 AM – 11.30 AM

CHEMISTRY
HIGHER

Fill in these boxes and read what is printed below.

Full name of centre

Town

Forename(s)

Surname

Date of birth
Day Month Year

Scottish candidate number

Number of seat

Reference may be made to the Chemistry Higher and Advanced Higher Data Booklet (1999 edition).

SECTION A—Questions 1–40 (40 marks)

Instructions for completion of **Section A** are given on page two.

For this section of the examination you must use an **HB pencil**.

SECTION B (60 marks)

1   All questions should be attempted.

2   The questions may be answered in any order but all answers are to be written in the spaces provided in this answer book, **and must be written clearly and legibly in ink**.

3   Rough work, if any should be necessary, should be written in this book and then scored through when the fair copy has been written. If further space is required, a supplementary sheet for rough work may be obtained from the invigilator.

4   Additional space for answers will be found at the end of the book. If further space is required, supplementary sheets may be obtained from the invigilator and should be inserted inside the **front cover of this book**.

5   The size of the space provided for an answer should not be taken as an indication of how much to write. It is not necessary to use all the space.

6   Before leaving the examination room you must give this book to the invigilator. If you do not, you may lose all the marks for this paper.

SCOTTISH
QUALIFICATIONS
AUTHORITY

©

SECTION A

Read carefully

1 Check that the answer sheet provided is for **Chemistry Higher (Section A)**.

2 For this section of the examination you must use an **HB pencil** and, where necessary, an eraser.

3 Check that the answer sheet you have been given has **your name**, **date of birth**, **SCN** (Scottish Candidate Number) and **Centre Name** printed on it.

Do not change any of these details.

4 If any of this information is wrong, tell the Invigilator immediately.

5 If this information is correct, **print** your name and seat number in the boxes provided.

6 The answer to each question is **either** A, B, C or D. Decide what your answer is, then, using your pencil, put a horizontal line in the space provided (see sample question below).

7 There is **only one correct** answer to each question.

8 Any rough working should be done on the question paper or the rough working sheet, **not** on your answer sheet.

9 At the end of the exam, put the **answer sheet for Section A inside the front cover of your answer book**.

Sample Question

To show that the ink in a ball-pen consists of a mixture of dyes, the method of separation would be

        A chromatography

        B fractional distillation

        C fractional crystallisation

        D filtration.

The correct answer is **A**—chromatography. The answer **A** has been clearly marked in **pencil** with a horizontal line (see below).

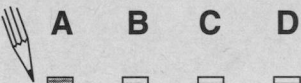

Changing an answer

If you decide to change your answer, carefully erase your first answer and using your pencil, fill in the answer you want. The answer below has been changed to **D**.

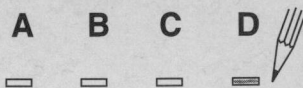

1. A negatively charged particle with electronic arrangement 2, 8 could be

   A a fluoride ion

   B a sodium atom

   C an aluminium ion

   D a neon atom.

2. Which of the following exists as diatomic molecules?

   A Helium

   B Methane

   C Carbon monoxide

   D Sodium chloride

3. When an atom **X** of an element in Group 1 reacts to become **X$^+$**

   A the mass number of **X** increases

   B the charge of the nucleus increases

   C the atomic number of **X** decreases

   D the number of filled energy levels decreases.

4. A mixture of magnesium chloride and magnesium sulphate is known to contain 0·6 mol of chloride ions and 0·2 mol of sulphate ions.

   What is the number of moles of magnesium ions present?

   A 0·4

   B 0·5

   C 0·8

   D 1·0

5. When a gas was bubbled through dilute hydrochloric acid, the pH increased.

   The gas could have been

   A ammonia

   B hydrogen

   C methane

   D sulphur dioxide.

6. The graph below shows the change in the concentration of a reactant with time for a given chemical reaction.

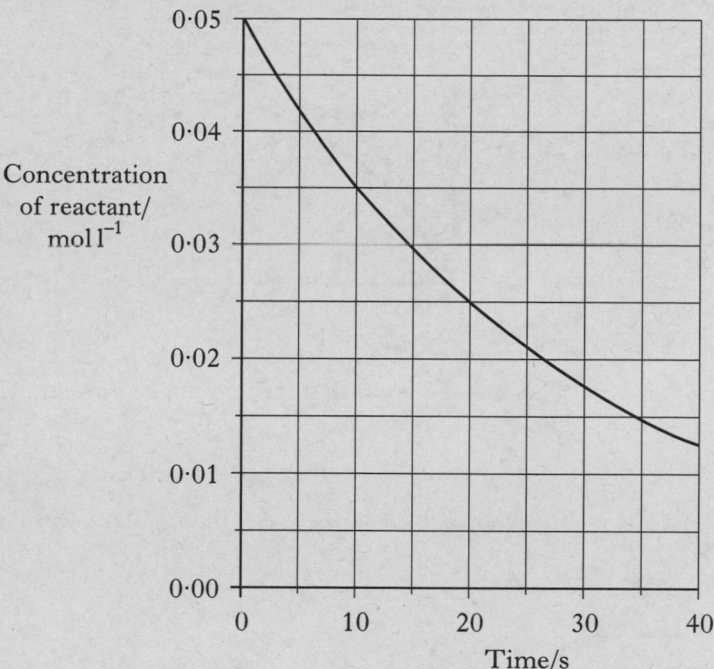

   What is the average rate of this reaction, in $mol\,l^{-1}\,s^{-1}$, between 10 and 20 s?

   A $1·0 \times 10^{-2}$

   B $1·0 \times 10^{-3}$

   C $1·5 \times 10^{-2}$

   D $1·5 \times 10^{-3}$

7. A small increase in temperature results in a large increase in rate of reaction.

   The **main** reason for this is that

   A more collisions are taking place

   B the enthalpy change is lowered

   C the activation energy is lowered

   D many more particles have energy greater than the activation energy.

   [Turn over

8. When copper carbonate reacts with excess acid, carbon dioxide is produced. The curves shown were obtained under two different conditions.

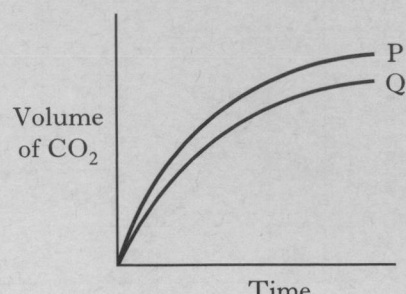

The change from **P** to **Q** could be brought about by

A increasing the concentration of the acid

B decreasing the mass of copper carbonate

C decreasing the particle size of the copper carbonate

D adding a catalyst.

9. $63 \cdot 5$ g of copper is added to 1 litre of $1 \text{ mol } l^{-1}$ silver(I) nitrate solution.

Which of the following statements is always true for this reaction?

A The resulting solution is colourless.

B All the copper dissolves.

C $63 \cdot 5$ g of silver is formed.

D One mole of silver is formed.

10. A potential energy diagram is shown.

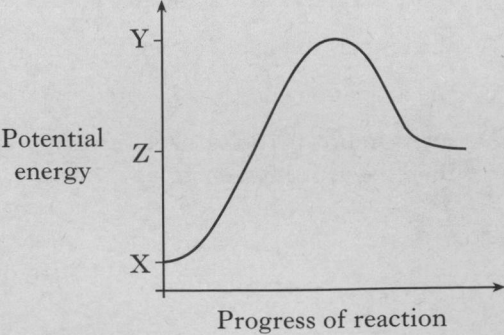

What is the activation energy (E_A) for the forward reaction?

A Y

B Z – X

C Y – X

D Y – Z

11. A group of students added 6 g of ammonium chloride crystals to 200 cm^3 of water at a temperature of $25\,°C$.

The enthalpy of solution of ammonium chloride is $+13 \cdot 6 \text{ kJ mol}^{-1}$.

After dissolving the crystals, the temperature of the solution would most likely be

A $23\,°C$

B $25\,°C$

C $27\,°C$

D $30\,°C$.

12. The spike graph shows the variation in successive ionisation energies of an element, **Z**.

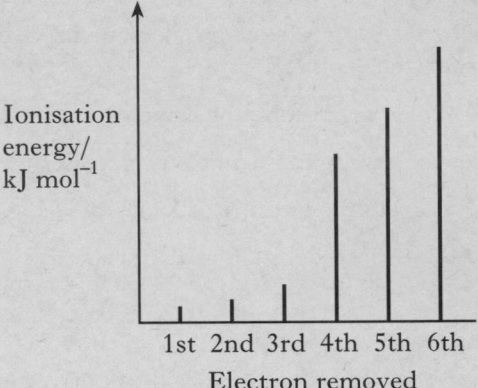

In which group of the Periodic Table is element **Z**?

A 1

B 3

C 4

D 6

13. Which equation represents the second ionisation energy of magnesium?

A $Mg^+(g) \rightarrow Mg^{2+}(g) + e^-$

B $Mg(g) \rightarrow Mg^{2+}(g) + 2e^-$

C $Mg(s) \rightarrow Mg^{2+}(g) + 2e^-$

D $Mg^+(s) \rightarrow Mg^{2+}(s) + e^-$

14. Carbon dioxide is a gas at room temperature while silicon dioxide is a solid because

A van der Waals' forces are much weaker than covalent bonds

B carbon dioxide contains double covalent bonds and silicon dioxide contains single covalent bonds

C carbon-oxygen bonds are less polar than silicon-oxygen bonds

D the relative formula mass of carbon dioxide is less than that of silicon dioxide.

15. Which of the following contains the same number of atoms as 16 g of helium?

A 16 g of methane

B 16 g of oxygen

C 17 g of ammonia

D 20 g of argon

16. A one carat diamond used in a ring contained 1×10^{22} carbon atoms.

What is the approximate mass of the diamond?

A 0·1 g

B 0·2 g

C 1·0 g

D 1·2 g

17. $C_2H_6(g) + 3\frac{1}{2}O_2(g) \rightarrow 2CO_2(g) + 3H_2O(\ell)$

When 20 cm^3 of ethane was sparked with 100 cm^3 of oxygen, what was the final volume of gases?

All volumes were measured at atmospheric pressure and room temperature.

A 40 cm^3

B 70 cm^3

C 100 cm^3

D 130 cm^3

18. During the manufacture of petrol, the process of reforming is used to produce a petrol with a higher percentage of molecules which are

A aromatic

B larger

C unbranched

D unsaturated.

19. An ester has the following structural formula:

$CH_3CH_2CH_2COOCH_2CH_3$

The name of this ester is

A propyl propanoate

B ethyl butanoate

C butyl ethanoate

D ethyl propanoate.

20. Which of the following compounds has isomeric forms?

A C_2H_3Cl

B C_2H_5Cl

C C_2HCl_3

D $C_2H_4Cl_2$

21. Which of the following structural formulae represents a primary alcohol?

A $CH_3 - CH_2 - CH_2 - \underset{\underset{\displaystyle OH}{|}}{\overset{\overset{\displaystyle H}{|}}{C}} - CH_3$

B $CH_3 - CH_2 - \underset{\underset{\displaystyle OH}{|}}{\overset{\overset{\displaystyle H}{|}}{C}} - CH_2 - CH_3$

C $CH_3 - \underset{\underset{\displaystyle OH}{|}}{\overset{\overset{\displaystyle CH_3}{|}}{C}} - CH_2 - CH_3$

D $CH_3 - \underset{\underset{\displaystyle CH_3}{|}}{\overset{\overset{\displaystyle CH_3}{|}}{C}} - CH_2 - OH$

[Turn over

22. What product(s) would be expected on dehydrating the following alcohol?

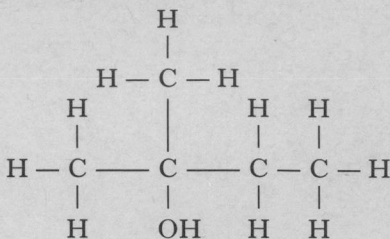

A   2-methylbut-2-ene only

B   2-methylbut-2-ene and
2-methylbut-1-ene

C   2-methylbut-1-ene only

D   3-methylbut-1-ene and
2-methylbut-1-ene

23. Which of the following statements about benzene is correct?

A   Benzene is an isomer of cyclohexane.

B   Benzene reacts with bromine solution as if it is unsaturated.

C   The ratio of carbon to hydrogen atoms in benzene is the same as in ethyne.

D   Benzene undergoes addition reactions more readily then hexene.

24. Which of the following consumer products is **least** likely to contain esters?

A   Flavourings

B   Perfumes

C   Solvents

D   Toothpastes

25. Which of the following types of bond is broken during hydrolysis of a polyamide?

A $C-N$

B $C=O$

C $N-H$

D $C-C$

26. Synthesis gas can be made by

A   reacting ethene gas with steam

B   burning carbon in excess air

C   burning methane gas in excess air

D   reacting methane gas with steam.

27. Some recently developed polymers have unusual properties.

Which polymer is soluble in water?

A   Poly(ethyne)

B   Poly(ethenol)

C   Biopol

D   Kevlar

28. The flow diagram shows two steps in the preparation of a feedstock for use in making a plastic.

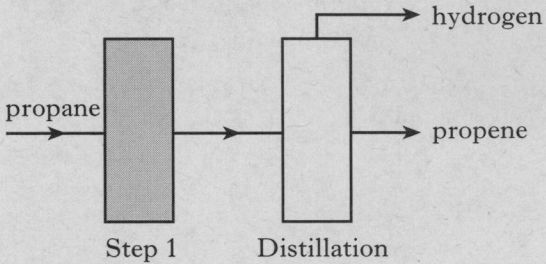

The reaction taking place during step 1 is an example of

A   cracking

B   oxidation

C   dehydration

D   hydrogenation.

29. What type of reaction is involved in the conversion of vegetable oils into "hardened" fats?

A   Condensation

B   Hydration

C   Hydrogenation

D   Polymerisation

30. Which of the following could be a monomer for the formation of a protein?

A $HOOC - CH_2 - COOH$

B $HOCH_2 - CHOH - CH_2OH$

C
$$\begin{array}{c} NH_2 \\ | \\ CH_3 - CH - COOH \end{array}$$

D
$$\begin{array}{c} OH \\ | \\ CH_3 - CH - COOH \end{array}$$

31. Consider the reaction pathways shown below.

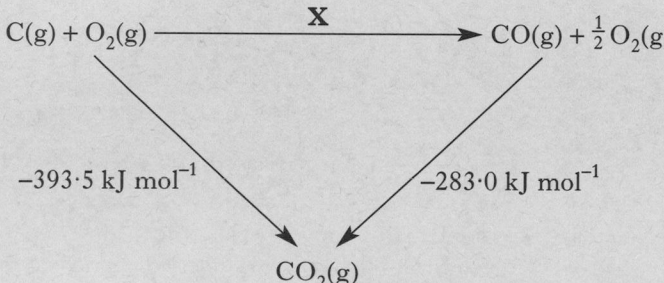

According to Hess's Law, the enthalpy change for reaction **X** is

A $+110 \cdot 5 \text{ kJ mol}^{-1}$

B $-110 \cdot 5 \text{ kJ mol}^{-1}$

C $-676 \cdot 5 \text{ kJ mol}^{-1}$

D $+676 \cdot 5 \text{ kJ mol}^{-1}$.

32. In a reversible reaction, equilibrium is reached when

A molecules of reactants cease to change into molecules of products

B the concentrations of reactants and products are equal

C the concentrations of reactants and products are constant

D the activation energy of the forward reaction is equal to that of the reverse reaction.

33. Gaseous iodine and hydrogen were reacted together in a sealed container.

$$I_2(g) + H_2(g) \rightleftharpoons 2HI(g)$$

Which of the following graphs shows the pressure inside the vessel as the reaction proceeds at constant temperature?

A
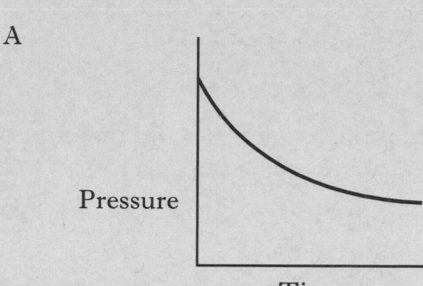

B
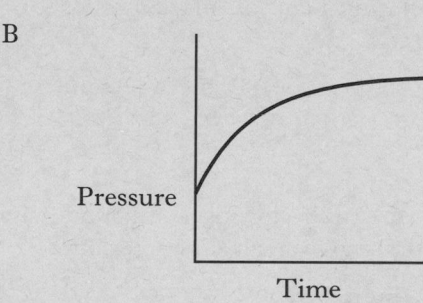

C
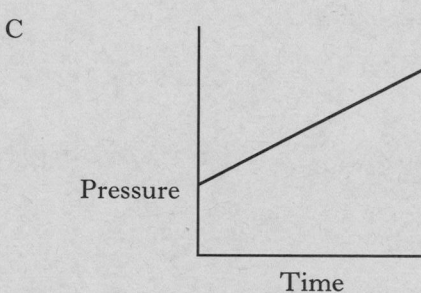

D
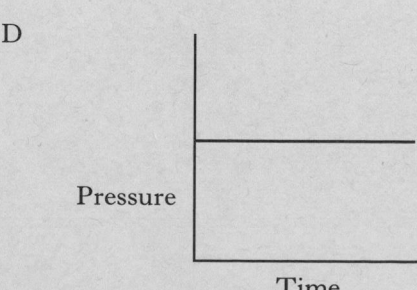

[Turn over

34. Which of the following is the same for equal volumes of $0.1\ mol\ l^{-1}$ solutions of sodium hydroxide and ammonia?

  A The pH of solution

  B The mass of solute present

  C The conductivity of solution

  D The number of moles of hydrochloric acid needed for neutralisation

35. Which of the following would cause the pH of an acid solution to be changed from 2 to 4?

  A Diluting $100\ cm^3$ of solution to $200\ cm^3$ with water

  B Evaporating $100\ cm^3$ of solution until $50\ cm^3$ remain

  C Diluting $1\ cm^3$ of solution to $100\ cm^3$ with water

  D Evaporating $100\ cm^3$ of solution until $1\ cm^3$ remains

36. Which of the following compounds dissolves in water to give an alkaline solution?

  A Sodium nitrate

  B Potassium ethanoate

  C Ammonium chloride

  D Lithium sulphate

37. $HgCl_2(aq) + SnCl_2(aq) \rightarrow Hg\ (\ell) + SnCl_4(aq)$

  What ion is oxidised in the above redox reaction?

  A $Sn^{2+}(aq)$

  B $Sn^{4+}(aq)$

  C $Hg^{2+}(aq)$

  D $Cl^-(aq)$

38. The unlabelled line on the graph below was obtained by plotting the mass of copper metal deposited against charge passed during the electrolysis of a melt of copper(I) chloride.

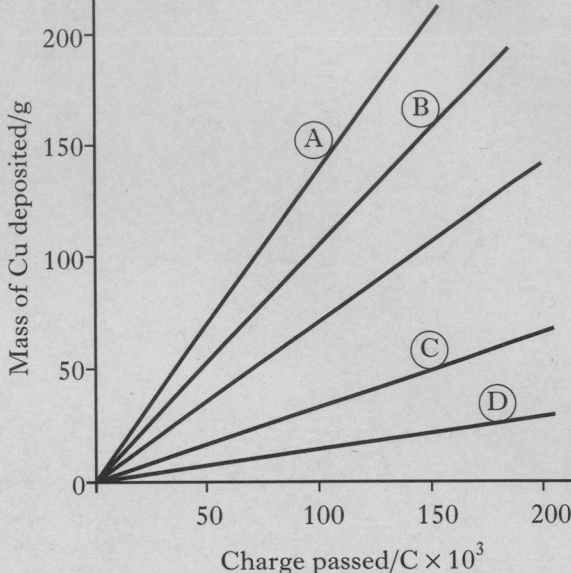

If a melt of copper(II) chloride was electrolysed, which line would correspond to the mass of copper metal deposited against charge passed?

39. ^{14}C has a half life of 5600 years. An analysis of charcoal from a wood fire shows that its ^{14}C content is 25% that of living wood.

  How many years have passed since the wood for the fire was cut?

  A 1400

  B 4200

  C 11 200

  D 16 800

40. A radioisotope of thorium forms protactinium-231 by beta-emission.

  What is the mass number of the radioisotope of thorium?

  A 230

  B 231

  C 232

  D 235

Candidates are reminded that the answer sheet MUST be returned INSIDE the front cover of this answer book.

Marks

SECTION B

All answers must be written clearly and legibly in ink.

1. The melting and boiling points and electrical conductivities of four substances are given in the table.

Substance	Melting point/°C	Boiling point/°C	Solid conducts electricity?	Melt conducts electricity?
A	92	190	no	no
B	1050	2500	yes	yes
C	773	1407	no	yes
D	1883	2503	no	no

Complete the table below by adding the appropriate letter for each type of bonding and structure.

Substance	Bonding and structure at room temperature
	covalent molecular
	covalent network
	ionic
	metallic

(2)

[Turn over

DO NOT WRITE IN THIS MARGIN

Marks

2. The elements in the second row of the Periodic Table are shown below.

Li	Be	B	C	N	O	F	Ne

(*a*) Why does the atomic size decrease crossing the period from lithium to neon? **1**

(*b*) Diamond and graphite are forms of carbon that exist as network solids.

Name a form of carbon that exists as discrete molecules. **1**

(*c*) Use the electronegativity values to explain why nitrogen chloride contains pure covalent bonds. **1**

(3)

Marks

3. Catalytic converters in car exhaust systems convert poisonous gases into less harmful gases.

(*a*) Two less harmful gases are formed when nitrogen monoxide reacts with carbon monoxide.

Name the **two** gases produced.

1

(*b*) The catalyst is made up of the metals platinum, palladium and rhodium.

Explain what happens to molecules in the exhaust gas during their catalytic conversion to less harmful gases.

You may wish to draw labelled diagrams.

2

(3)

[Turn over

Marks

4. The enthalpy of combustion of methanol (CH_3OH) can be determined from measurements using the apparatus shown.

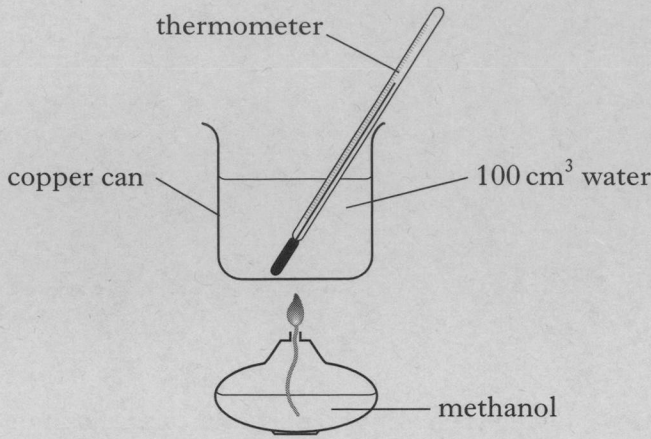

thermometer

copper can — 100 cm³ water

methanol

(*a*) In an experiment, the following results were obtained.

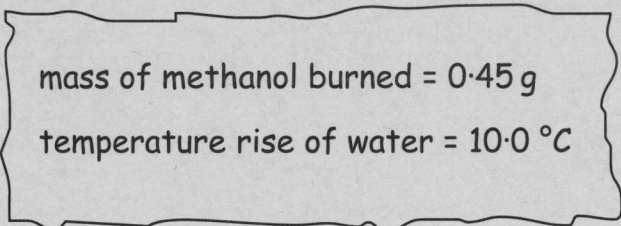

mass of methanol burned = 0·45 g

temperature rise of water = 10·0 °C

Use these results to calculate the enthalpy of combustion, in $kJ\ mol^{-1}$, of methanol.

Show your working clearly.

3

DO NOT WRITE IN THIS MARGIN

Marks

4. (continued)

(*b*) Suggest **two** reasons why the experimental value for the enthalpy of combustion of methanol is different from the value given on page 9 of the data booklet. **2**

(5)

[Turn over

Marks

5. Glycerol trinitrate is an explosive with the structure shown.

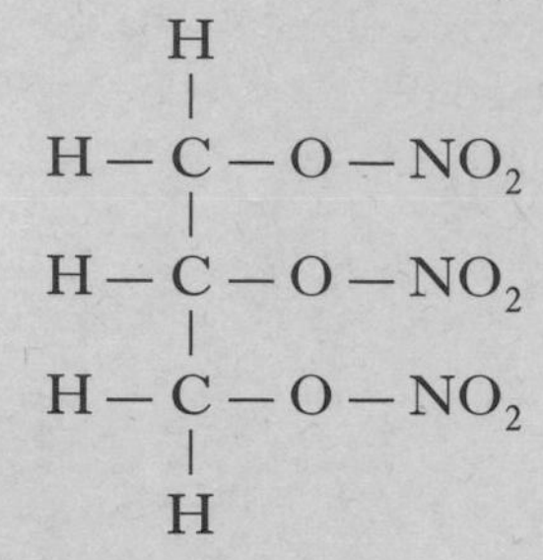

$$
\begin{array}{c}
H \\
| \\
H - C - O - NO_2 \\
| \\
H - C - O - NO_2 \\
| \\
H - C - O - NO_2 \\
| \\
H
\end{array}
$$

(*a*) Glycerol trinitrate is produced from glycerol.

    (i) Draw a structural formula for glycerol.

1

    (ii) Name a group of naturally occurring esters that can be hydrolysed to obtain glycerol.

1

(*b*) When exploded, glycerol trinitrate decomposes to give nitrogen, water, carbon dioxide and oxygen.

Balance the equation for this reaction.

$$C_3H_5N_3O_9(\ell) \rightarrow N_2(g) + H_2O(g) + CO_2(g) + O_2(g)$$

1

(3)

Marks

6. An enzyme found in potatoes can catalyse the decomposition of hydrogen peroxide.

$$2H_2O_2(aq) \quad \rightarrow \quad 2H_2O(\ell) \quad + \quad O_2(g)$$

The rate of the decomposition of hydrogen peroxide can be studied using the apparatus shown.

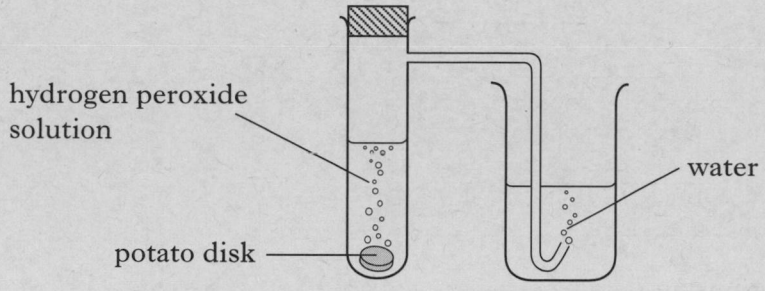

hydrogen peroxide
solution

water

potato disk

(a) Describe how this apparatus can be used to investigate the effect of temperature on the rate of decomposition of hydrogen peroxide.

2

(b) The graph shows how the rate of the enzyme catalysed reaction changes with temperature.

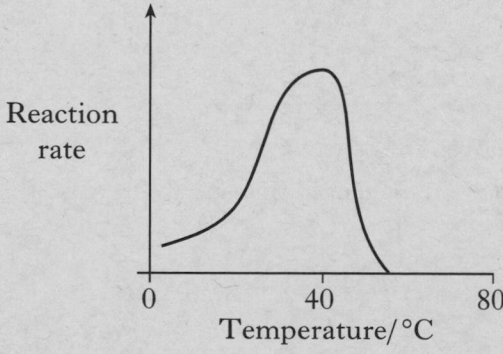

Reaction
rate

0   40   80
Temperature/°C

Why does the reaction rate decrease above the optimum temperature of 40 °C?

1

(3)

Marks

7. An industrial method for the production of ethanol is outlined in the flow diagram.

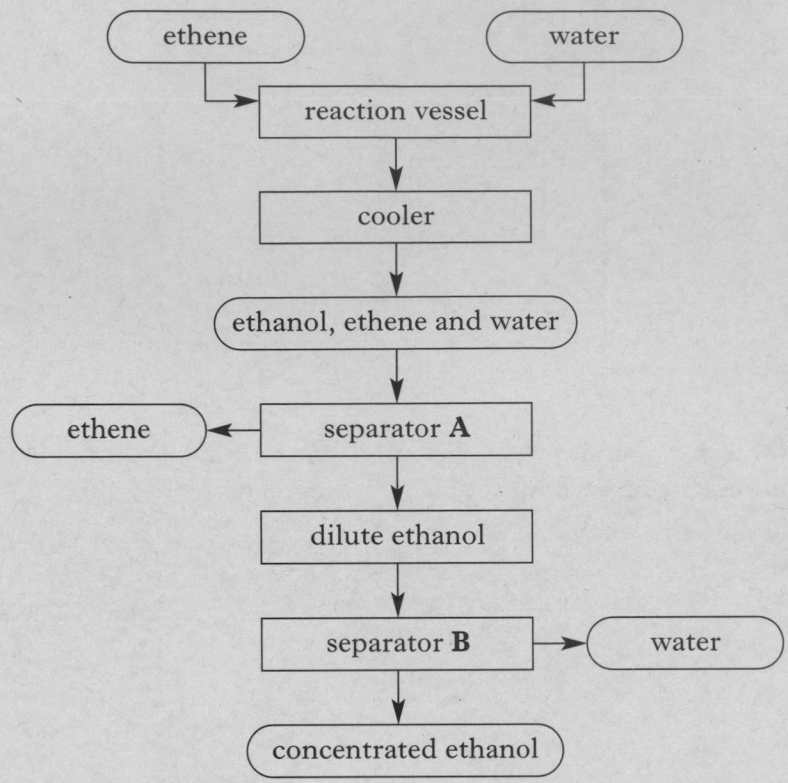

(a) The starting materials are ethene and water. Water is a raw material but ethene is not. Why is ethene **not** a raw material?

1

(b) (i) Unreacted ethene is removed in separator **A**.

Suggest how the separated ethene could be used to increase the efficiency of the overall process.

1

(ii) Name the process that takes place in separator **B**.

1

DO NOT WRITE IN THIS MARGIN

Marks

7. (continued)

(*c*) In the reaction vessel, ethanol is produced in an exothermic reaction.

$$C_2H_4(g) \quad + \quad H_2O(g) \quad \rightleftharpoons \quad C_2H_5OH(g)$$

(i) Name the type of chemical reaction that takes place in the reaction vessel. **1**

(ii) What would happen to the equilibrium position if the temperature inside the reaction vessel was increased? **1**

(iii) If 1·64 kg of ethanol (relative formula mass = 46) is produced from 10·0 kg of ethene (relative formula mass = 28), calculate the percentage yield of ethanol. **2**

(7)

Marks

8. "Self-test" kits can be used to check the quantity of alcohol present in a person's breath.

The person blows through a glass tube until a plastic bag at the end is completely filled.

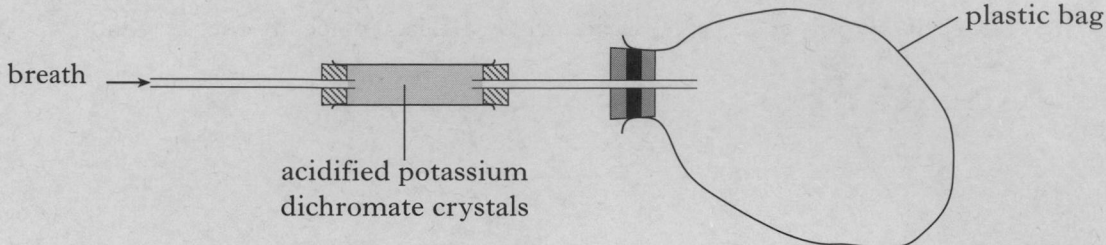

breath →

acidified potassium
dichromate crystals

plastic bag

The tube contains orange acidified potassium dichromate crystals that turn green when they react with ethanol. The chemical reaction causing the colour change is:

$$Cr_2O_7^{2-} + 14H^+ + 6e^- \longrightarrow 2Cr^{3+} + 7H_2O$$

orange   green

The more ethanol present in the person's breath, the further along the tube the green colour travels.

(*a*) What is the purpose of the plastic bag?

1

(*b*) Why are the potassium dichromate crystals acidified?

1

(*c*) Name a carbon compound formed by the reaction of ethanol with acidified potassium dichromate crystals.

1

(3)

Marks

9. Car tyres can be made from natural or synthetic rubber.

Natural rubber is a thermoplastic polymer. The monomer used for the polymerisation is called 2-methylbuta-1,3-diene.

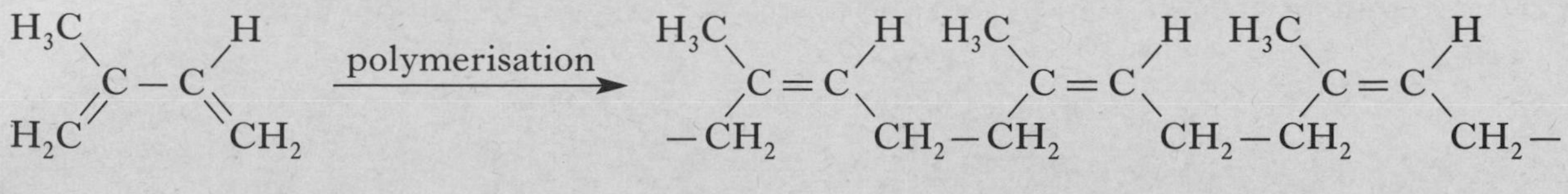

2-methylbuta-1,3-diene   poly(2-methylbuta-1,3-diene)

(*a*) Synthetic rubber is a similar polymer.

Part of the structure of synthetic rubber showing three monomer units linked together is shown.

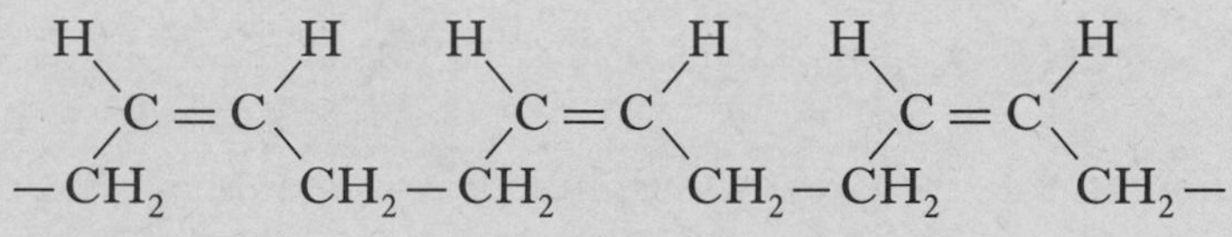

(i) Draw the monomer used to form synthetic rubber.

1

(ii) Name the type of polymerisation involved in the manufacture of synthetic rubber.

1

(*b*) In making tyre rubber, the polymer chains are cross-linked to give a three-dimensional structure by heating with sulphur and sulphur compounds.

In what way will the **properties** of the rubber be improved by cross-linking?

1

(3)

Marks

10. When copper is added to an acid, what happens depends on the acid and the conditions.

    (a) When copper reacts with dilute nitric acid, nitrate ions are reduced.

    Complete the ion-electron equation for the reduction of the nitrate ions.

    $$NO_3^-(aq) \qquad\qquad \rightarrow \quad NO(g)$$

    1

    (b) Copper reacts with hot, concentrated sulphuric acid to produce sulphur dioxide.

    $$Cu \;+\; 2H_2SO_4 \;\rightarrow\; CuSO_4 \;+\; SO_2 \;+\; 2H_2O$$

    Calculate the volume, in litres, of sulphur dioxide gas that would be produced when 10 g of copper reacts with excess concentrated sulphuric acid.

    (Take the molar volume of sulphur dioxide to be 24 litres mol^{-1}.)

    Show your working clearly.

    2

    (3)

Marks

11. (*a*) State the pH of $0.01 \ mol \, l^{-1}$ hydrochloric acid.

1

(*b*) The pH values of $0.01 \ mol \, l^{-1}$ ethanoic acid and $0.01 \ mol \, l^{-1}$ sulphuric acid were measured.

The results were compared to the pH of $0.01 \ mol \, l^{-1}$ hydrochloric acid.

Circle the words in the table that show the correct comparisons.

Solution	pH compared to hydrochloric acid
ethanoic acid	higher / lower / the same
sulphuric acid	higher / lower / the same

2

(*c*) In pure ethanoic acid, the molecules are held together in pairs.

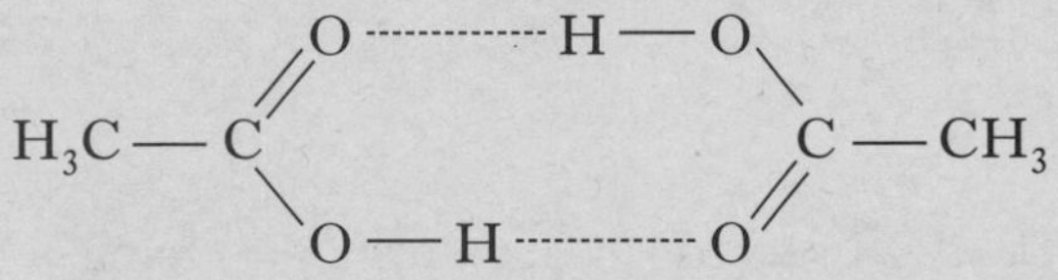

Explain why this happens.

In your answer you should name the type of intermolecular forces involved in pure ethanoic acid and explain how they arise.

2

(5)

[Turn over

Marks

12. The Swiss-born Russian chemist Germain Henri Hess first put forward his law in 1840.

(*a*)　(i)　Hess's Law can be verified using the reactions summarised below.

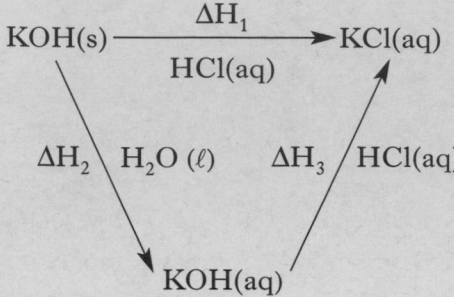

Complete the list of measurements that would have to be made in order to determine ΔH_3.

1　volume of potassium hydroxide solution

2

3

4

5

2

(ii)　Hess's Law can be used to obtain enthalpy changes for reactions that cannot be measured directly.

Use the following enthalpy changes

$$KClO_3(s) + 3Mg(s) \rightarrow KCl(s) + 3MgO(s) \qquad \Delta H = -1852\,kJ\,mol^{-1}$$
$$K(s) + \tfrac{1}{2}Cl_2(g) \rightarrow KCl(s) \qquad \Delta H = -437\,kJ\,mol^{-1}$$
$$Mg(s) + \tfrac{1}{2}O_2(g) \rightarrow MgO(s) \qquad \Delta H = -602\,kJ\,mol^{-1}$$

to calculate the enthalpy change in $kJ\,mol^{-1}$, for the reaction:

$$K(s) + \tfrac{1}{2}Cl_2(g) + 1\tfrac{1}{2}O_2(g) \rightarrow KClO_3(s)$$

2

Marks

12. **(continued)**

(*b*) The enthalpy change for the reaction

$$K(s) \; + \; \tfrac{1}{2}Cl_2(g) \; + \; 1\tfrac{1}{2}O_2(g) \; \rightarrow \; KClO_3(s)$$

is an example of an enthalpy of formation.

The enthalpy of formation of a compound can be defined as the enthalpy change for the formation of one mole of a compound from its elements as they exist at room temperature.

Write the equation, including state symbols, corresponding to the enthalpy of formation of sodium oxide (Na_2O).

1

(5)

[Turn over

Marks

13. When a mixture of solid sodium hydroxide and solid sodium ethanoate is heated, methane gas and solid sodium carbonate are produced.

$$NaOH(s) \ + \ CH_3COONa(s) \ \rightarrow \ CH_4(g) \ + \ Na_2CO_3(s)$$

(a) Draw a diagram of an apparatus which could be used for this reaction showing how the methane gas can be collected and its volume measured.

2

(b) Name the gas produced when sodium propanoate is heated with solid sodium hydroxide.

1

(3)

Marks

14. Americium-241 is an alpha-emitting radioisotope that is used in a popular type of smoke detector.

(*a*)   (i) Write a balanced nuclear equation for the alpha decay of americium-241 (atomic number 95).

1

    (ii) Suggest a reason why this type of smoke detector is not regarded as a health hazard, even though it contains a radioactive source.

1

(*b*) The americium-241 is present in the smoke detector as the metal oxide.

If the smoke detector contains $0.00025\,g$ of americium-241 oxide, AmO_2, calculate the mass present, in grams, of americium-241.

1

(3)

[Turn over

Marks

15. Chlorine can be produced commercially from concentrated sodium chloride solution in a membrane cell.

Only sodium ions can pass through the membrane. These ions move in the direction shown in the diagram.

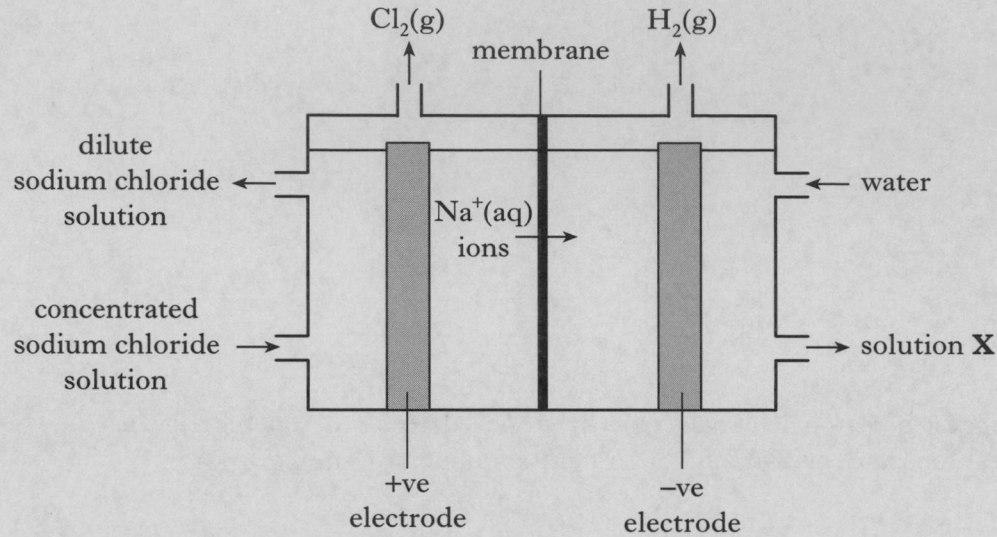

The reactions at each electrode are:

+ve electrode: $2Cl^-(aq) \rightarrow Cl_2(g) + 2e^-$

−ve electrode: $2H_2O(\ell) + 2e^- \rightarrow H_2(g) + 2OH^-(aq)$

(a)   Write the overall redox equation for the reaction in the membrane cell.

1

(b)   (i)   Name solution **X**.

1

(ii)   Hydrogen gas is produced at the negative electrode in the membrane cell.
Suggest why this gas could be a valuable resource in the future.

1

Marks

15.   (continued)

(c)   Calculate the mass of chlorine, in kilograms, produced in a membrane cell using a current of 80 000 A for 10 hours.

Show your working clearly.

3

(6)

[Turn over

Marks

16. Infrared spectroscopy can be used to help identify the bonds which are present in an organic molecule.

Different bonds absorb infrared radiation of different wave numbers.

The table below shows the range of wave numbers of infrared radiation absorbed by the bonds indicated with thicker lines.

Bond	Wave number range/cm^{-1}
O — H	3650 – 3590
C ≡ C — H	3300
C — C — H	2962 – 2853
C ≡ C	2260 – 2100
C — O	1150 – 1070

The infrared spectrum and full structural formula for compound ① are shown.

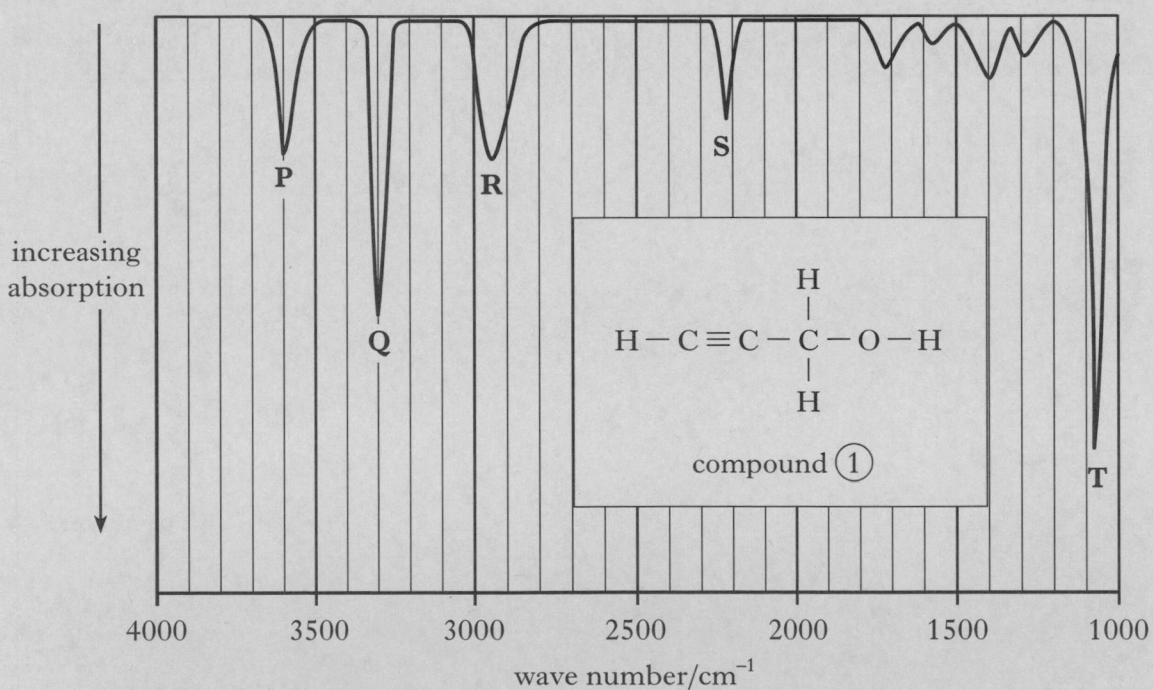

(a) In the above full structural formula for compound ①, use an arrow to indicate the bond that could be responsible for the absorption at **T** in its infrared spectrum.

1

Marks

16. (continued)

(*b*) A series of reactions is carried out starting with compound ①.

$$\text{compound} \; ① \; \xrightarrow{\;1 \text{ mol } H_2\;} \; \text{compound} \; ② \; \xrightarrow{\;1 \text{ mol } H_2\;} \; \text{compound} \; ③$$

(i) Give the letters for the **two** absorptions which would **not** appear in the infrared spectrum for compound ②.

1

(ii) Name compound ③.

1

(3)

[END OF QUESTION PAPER]

ADDITIONAL SPACE FOR ANSWERS

ADDITIONAL SPACE FOR ANSWERS

ADDITIONAL SPACE FOR ANSWERS

[BLANK PAGE]

FOR OFFICIAL USE

Total
Section B

X012/301

NATIONAL
QUALIFICATIONS
2007

TUESDAY, 29 MAY
9.00 AM – 11.30 AM

CHEMISTRY
HIGHER

Fill in these boxes and read what is printed below.

Full name of centre

Town

Forename(s)

Surname

Date of birth
Day Month Year

Scottish candidate number

Number of seat

Reference may be made to the Chemistry Higher and Advanced Higher Data Booklet.

SECTION A—Questions 1–40 (40 marks)

Instructions for completion of **Section A** are given on page two.

For this section of the examination you must use an **HB pencil**.

SECTION B (60 marks)

1   All questions should be attempted.

2   The questions may be answered in any order but all answers are to be written in the spaces provided in this answer book, **and must be written clearly and legibly in ink**.

3   Rough work, if any should be necessary, should be written in this book and then scored through when the fair copy has been written. If further space is required, a supplementary sheet for rough work may be obtained from the invigilator.

4   Additional space for answers will be found at the end of the book. If further space is required, supplementary sheets may be obtained from the invigilator and should be inserted inside the **front** cover of this book.

5   The size of the space provided for an answer should not be taken as an indication of how much to write. It is not necessary to use all the space.

6   Before leaving the examination room you must give this book to the invigilator. If you do not, you may lose all the marks for this paper.

SCOTTISH
QUALIFICATIONS
AUTHORITY

SECTION A

Read carefully

1 Check that the answer sheet provided is for **Chemistry Higher (Section A)**.

2 For this section of the examination you must use an **HB pencil** and, where necessary, an eraser.

3 Check that the answer sheet you have been given has **your name**, **date of birth**, **SCN** (Scottish Candidate Number) and **Centre Name** printed on it.

   Do not change any of these details.

4 If any of this information is wrong, tell the Invigilator immediately.

5 If this information is correct, **print** your name and seat number in the boxes provided.

6 The answer to each question is **either** A, B, C or D. Decide what your answer is, then, using your pencil, put a horizontal line in the space provided (see sample question below).

7 There is **only one correct** answer to each question.

8 Any rough working should be done on the question paper or the rough working sheet, **not** on your answer sheet.

9 At the end of the exam, put the **answer sheet for Section A inside the front cover of your answer book**.

Sample Question

To show that the ink in a ball-pen consists of a mixture of dyes, the method of separation would be

      A chromatography

      B fractional distillation

      C fractional crystallisation

      D filtration.

The correct answer is **A**—chromatography. The answer **A** has been clearly marked in **pencil** with a horizontal line (see below).

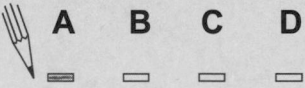

Changing an answer

If you decide to change your answer, carefully erase your first answer and using your pencil, fill in the answer you want. The answer below has been changed to **D**.

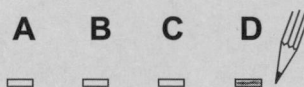

1. Which of the following compounds contains **both** a halide ion and a transition metal ion?

   A   Iron oxide

   B   Silver bromide

   C   Potassium permanganate

   D   Copper iodate

2. Which of the following substances is a non-conductor but becomes a good conductor on melting?

   A   Solid potassium fluoride

   B   Solid argon

   C   Solid potassium

   D   Solid tetrachloromethane

3. Particles with the same electron arrangement are said to be isoelectronic.

   Which of the following compounds contains ions which are isoelectronic?

   A Na_2S

   B $MgCl_2$

   C   KBr

   D $CaCl_2$

4. The graph shows the variation of concentration of a reactant with time as a reaction proceeds.

   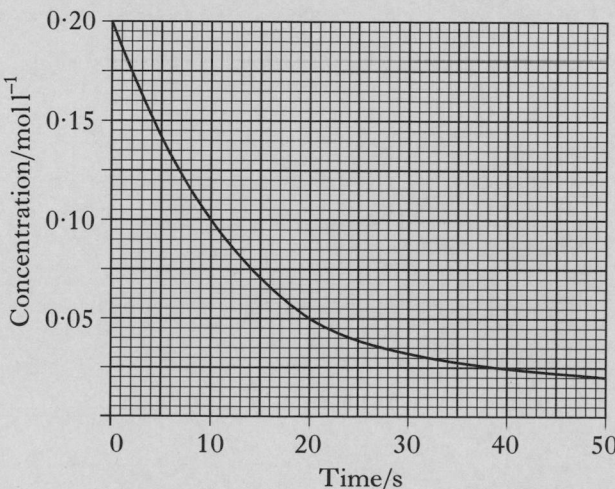

   What is the average reaction rate during the first 20 s?

   A $0.0025 \ mol \ l^{-1} s^{-1}$

   B $0.0050 \ mol \ l^{-1} s^{-1}$

   C $0.0075 \ mol \ l^{-1} s^{-1}$

   D $0.0150 \ mol \ l^{-1} s^{-1}$

5. The potential energy diagram below refers to the reversible reaction involving reactants **R** and products **P**.

   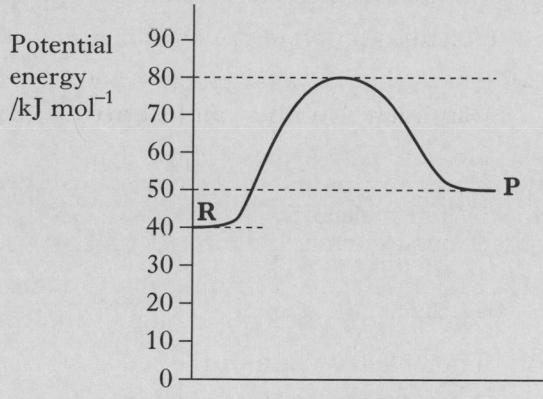

   What is the enthalpy change, in $kJ \ mol^{-1}$, for the reverse reaction $\mathbf{P} \rightarrow \mathbf{R}$?

   A   + 30

   B   + 10

   C   − 10

   D   − 40

6. The enthalpy of neutralisation in an acid/alkali reaction is **always** the energy released in

   A   the formation of one mole of salt

   B   the formation of one mole of water

   C   the neutralisation of one mole of acid

   D   the neutralisation of one mole of alkali.

7. Which equation represents the first ionisation energy of a diatomic element, X_2?

   A $\frac{1}{2}X_2(s) \rightarrow X^+(g)$

   B $\frac{1}{2}X_2(g) \rightarrow X^-(g)$

   C $X(g) \rightarrow X^+(g)$

   D $X(s) \rightarrow X^-(g)$

8. Which of the following chlorides is likely to have **least** ionic character?

   A $BeCl_2$

   B $CaCl_2$

   C   LiCl

   D   CsCl

[Turn over

9. Which of the following chlorides is most likely to be soluble in tetrachloromethane, CCl_4?

   A Barium chloride

   B Caesium chloride

   C Calcium chloride

   D Phosphorus chloride

10. Which of the following compounds exists as discrete molecules?

    A Sulphur dioxide

    B Silicon dioxide

    C Aluminium oxide

    D Iron(II) oxide

11. An element (melting point above 3000 °C) forms an oxide which is a gas at room temperature.

    Which type of bonding is likely to be present in the element?

    A Metallic

    B Polar covalent

    C Non-polar covalent

    D Ionic

12. Which of the following compounds has polar molecules?

    A CO_2

    B NH_3

    C CCl_4

    D CH_4

13. How many moles of oxygen atoms are in 0·5 mol of carbon dioxide?

    A 0·25

    B 0·5

    C 1

    D 2

14. A fullerene molecule consists of 60 carbon atoms.

    Approximately how many such molecules are present in 12 g of this type of carbon?

    A $1·0 \times 10^{22}$

    B $1·2 \times 10^{23}$

    C $6·0 \times 10^{23}$

    D $3·6 \times 10^{25}$

15. Avogadro's Constant is the same as the number of

    A molecules in 16·0 g of oxygen

    B atoms in 20·2 g of neon

    C formula units in 20·0 g of sodium hydroxide

    D ions in 58·5 g of sodium chloride.

16. The equation for the complete combustion of propane is:

    $$C_3H_8(g) + 5O_2(g) \rightarrow 3CO_2(g) + 4H_2O(\ell)$$

    30 cm^3 of propane is mixed with 200 cm^3 of oxygen and the mixture is ignited.

    What is the volume of the resulting gas mixture? (All volumes are measured at the same temperature and pressure.)

    A 90 cm^3

    B 120 cm^3

    C 140 cm^3

    D 210 cm^3

17. A mixture of carbon monoxide and hydrogen can be converted into water and a mixture of hydrocarbons.

    $$nCO + (2n+1)H_2 \rightarrow nH_2O + hydrocarbons$$

    What is the general formula for the hydrocarbons produced?

    A C_nH_{2n-2}

    B C_nH_{2n}

    C C_nH_{2n+1}

    D C_nH_{2n+2}

18. Chemical processes are used to produce a petrol that burns more efficiently.

Which of the following types of hydrocarbon does **not** improve the burning efficiency of petrol?

A   straight-chain alkanes

B   branched-chain alkanes

C   cycloalkanes

D   aromatics

19. Which of the following is an aldehyde?

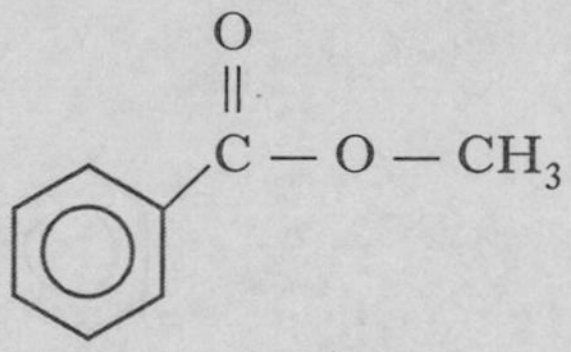

20. The dehydration of butan-2-ol can produce two isomeric alkenes, but-1-ene and but-2-ene.

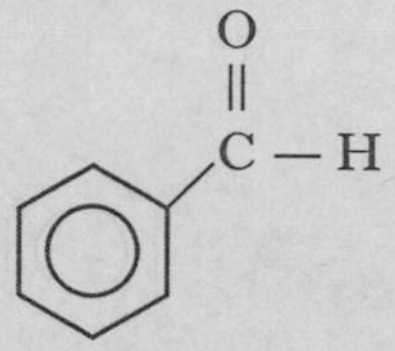

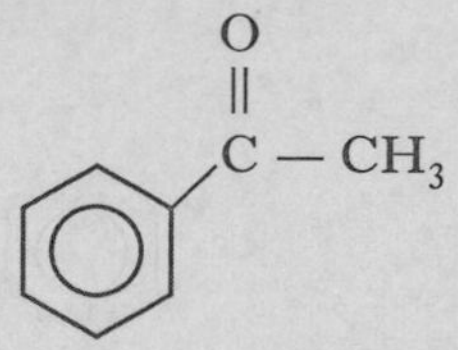

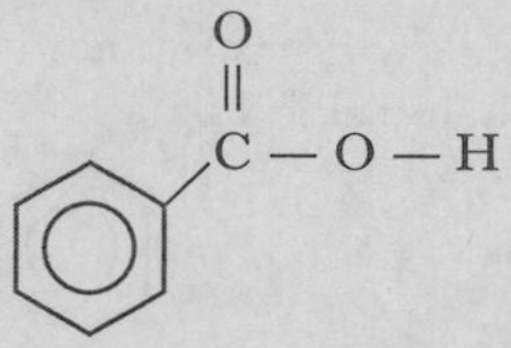

Which of the following alkanols can similarly produce, on dehydration, a pair of isomeric alkenes?

A   Propan-2-ol

B   Pentan-3-ol

C   Hexan-3-ol

D   Heptan-4-ol

21. Which of the following reactions can be classified as reduction?

A CH_3CH_2OH $\rightarrow CH_3COOH$

B $CH_3CH(OH)CH_3 \rightarrow CH_3COCH_3$

C $CH_3CH_2COCH_3$ $\rightarrow CH_3CH_2CH(OH)CH_3$

D CH_3CH_2CHO $\rightarrow CH_3CH_2COOH$

22. Which of the following compounds would react with sodium hydroxide solution to form a salt?

A CH_3CHO

B CH_3COOH

C CH_3COCH_3

D CH_3CH_2OH

23. The extensive use of which type of compound is thought to contribute significantly to the depletion of the ozone layer?

A   Oxides of carbon

B   Hydrocarbons

C   Oxides of sulphur

D   Chlorofluorocarbons

24. Propene is used in the manufacture of addition polymers.

What type of reaction is used to produce propene from propane.

A   Addition

B   Cracking

C   Hydrogenation

D   Oxidation

25. Cured polyester resins

A   are used as textile fibres

B   are long chain molecules

C   are formed by addition polymerisation

D   have a three-dimensional structure with cross linking.

[**Turn over**

26. What is the structural formula for glycerol?

A
$$CH_2OH$$
$$|$$
$$CH_2$$
$$|$$
$$CH_2OH$$

B
$$CH_2OH$$
$$|$$
$$CH_2OH$$

C
$$CH_2OH$$
$$|$$
$$CHOH$$
$$|$$
$$CH_2COOH$$

D
$$CH_2OH$$
$$|$$
$$CHOH$$
$$|$$
$$CH_2OH$$

27. The monomer units used to construct enzyme molecules are

A   alcohols

B   esters

C   amino acids

D   fatty acids.

28. In α-amino acids the amino group is on the carbon atom adjacent to the acid group.

Which of the following is an α-amino acid?

A $CH_3 - CH - COOH$
     $|$
     $CH_2 - NH_2$

B $CH_2 - CH - COOH$
     $|$ $|$
     SH NH_2

C NH_2

   COOH

D NH_2

   COOH

29. Which of the following compounds is **not** a raw material in the chemical industry?

A   Benzene

B   Water

C   Iron ore

D   Sodium chloride

30. $N_2(g) + 2O_2(g) \rightarrow 2NO_2(g)$ $\Delta H = +88\,kJ$
$N_2(g) + 2O_2(g) \rightarrow N_2O_4(g)$ $\Delta H = +10\,kJ$

The enthalpy change for the reaction

$$2NO_2(g) \rightarrow N_2O_4(g)$$

will be

A $+98\,kJ$

B $+78\,kJ$

C $-78\,kJ$

D $-98\,kJ$.

31. A catalyst is used in the Haber Process.

$$N_2(g) + 3H_2(g) \rightleftharpoons 2NH_3(g)$$

Which of the following best describes the action of the catalyst?

A   Increases the rate of the forward reaction only

B   Increases the rate of the reverse reaction only

C   Increases the rate of both the forward and reverse reactions

D   Changes the position of the equilibrium of the reaction

32. In which of the following systems will the equilibrium be **unaffected** by a change in pressure?

A $2NO_2(g) \rightleftharpoons N_2O_4(g)$

B $H_2(g) + I_2(g) \rightleftharpoons 2HI(g)$

C $N_2(g) + 3H_2(g) \rightleftharpoons 2NH_3(g)$

D $2NO(g) + O_2(g) \rightleftharpoons 2NO_2(g)$

33. On the structure shown, four hydrogen atoms have been replaced by letters **W**, **X**, **Y** and **Z**.

Which letter corresponds to the hydrogen atom which can ionise most easily in aqueous solution?

A   **W**

B   **X**

C   **Y**

D   **Z**

34. The concentration of $OH^-(aq)$ ions in a solution is $0.1 \, mol \, l^{-1}$.

What is the pH of the solution?

A   1

B   8

C   13

D   14

35. A lemon juice is found to have a pH of 3 and an apple juice a pH of 5.

From this information, the concentrations of $H^+(aq)$ ions in the lemon juice and apple juice are in the proportion (ratio)

A   100 : 1

B   1 : 100

C   20 : 1

D   3 : 5.

36. Which line in the table is correct for $0.1 \, mol \, l^{-1}$ sodium hydroxide solution compared with $0.1 \, mol \, l^{-1}$ ammonia solution?

	pH	Conductivity
A	higher	lower
B	higher	higher
C	lower	higher
D	lower	lower

37. The iodate ion, IO_3^-, can be converted to iodine.

Which is the correct ion-electron equation for the reaction?

A $2IO_3^-(aq) + 12H^+(aq) + 12e^- \rightarrow 2I^-(aq) + 6H_2O(\ell)$

B $IO_3^-(aq) + 6H^+(aq) + 7e^- \rightarrow I^-(aq) + 3H_2O(\ell)$

C $2IO_3^-(aq) + 12H^+(aq) + 11e^- \rightarrow I_2(aq) + 6H_2O(\ell)$

D $2IO_3^-(aq) + 12H^+(aq) + 10e^- \rightarrow I_2(aq) + 6H_2O(\ell)$

38. Which of the following is a redox reaction?

A $Mg + 2HCl \rightarrow MgCl_2 + H_2$

B $MgO + 2HCl \rightarrow MgCl_2 + H_2O$

C $MgCO_3 + 2HCl \rightarrow MgCl_2 + H_2O + CO_2$

D $Mg(OH)_2 + 2HCl \rightarrow MgCl_2 + 2H_2O$

39. Strontium-90 is a radioisotope.

What is the neutron to proton ratio in an atom of this isotope?

A   0.730

B   1.00

C   1.37

D   2.37

40. The diagram shows the paths of alpha, beta and gamma radiations as they pass through an electric field.

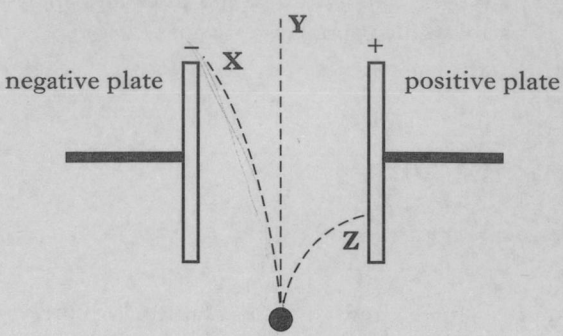

negative plate   positive plate

radioactive source

Which line in the table correctly identifies the types of radiation which follow paths **X**, **Y** and **Z**?

	Path X	Path Y	Path Z
A	gamma	beta	alpha
B	beta	gamma	alpha
C	beta	alpha	gamma
D	alpha	gamma	beta

Candidates are reminded that the answer sheet MUST be returned INSIDE the front cover of this answer book.

DO NOT WRITE IN THIS MARGIN

Marks

SECTION B

All answers must be written clearly and legibly in ink.

1. (*a*) Atoms of different elements have different attractions for bonded electrons.

What term is used as a measure of the attraction an atom involved in a bond has for the electrons of the bond? **1**

(*b*) Atoms of different elements are different sizes.

What is the trend in atomic size across the period from sodium to argon? **1**

(*c*) Atoms of different elements have different ionisation energies.

Explain clearly why the first ionisation energy of potassium is less than the first ionisation energy of sodium. **2**

(4)

[Turn over

Marks

2. Carbon compounds take part in a wide variety of chemical reactions.

(*a*)

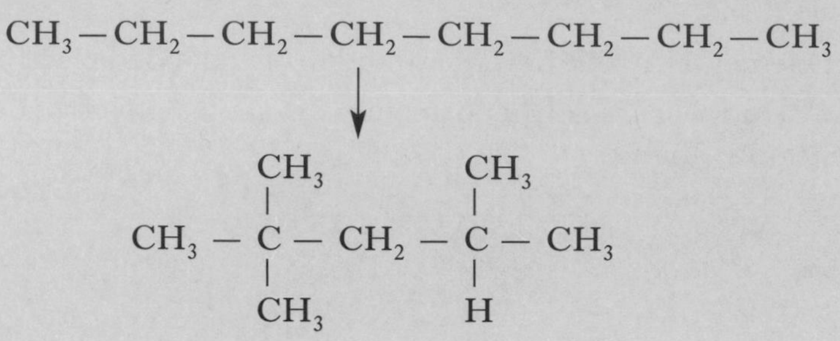

$$CH_3-CH_2-CH_2-CH_2-CH_2-CH_2-CH_2-CH_3$$

$$\downarrow$$

$$\underset{\underset{CH_3}{|}}{\overset{\overset{CH_3}{|}}{CH_3-C-CH_2-}}\underset{\overset{|}{H}}{\overset{\overset{CH_3}{|}}{C-CH_3}}$$

Name this type of chemical reaction.

1

(*b*) $$C_3H_6O \xrightarrow{\text{oxidation}} \text{propanoic acid}$$

Draw a structural formula for C_3H_6O.

1

(*c*) Kevlar is an aromatic polyamide made by condensation polymerisation.
Give **one** use for Kevlar.

1

(3)

Marks

3. Tritium, ^3_1H, is an isotope of hydrogen. It is formed in the upper atmosphere when neutrons from cosmic rays are captured by nitrogen atoms.

$$^{14}_{7}\text{N} \ + \ ^1_0\text{n} \ \rightarrow \ ^{12}_{6}\text{C} \ + \ ^3_1\text{H}$$

Tritium atoms then decay by beta-emission.

$$^3_1\text{H} \quad \rightarrow \quad \quad +$$

(a) Complete the nuclear equation above for the beta-decay of tritium atoms. **1**

(b) In the upper atmosphere, tritium atoms are present in some water molecules. Over the years, the concentration of tritium atoms in rain has remained fairly constant.

(i) Why does the concentration of tritium in rain remain fairly constant?

1

(ii) The concentration of tritium atoms in fallen rainwater is found to decrease over time. The age of any product made with water can be estimated by measuring the concentration of tritium atoms.

In a bottle of wine, the concentration of tritium atoms was found to be $\frac{1}{8}$ of the concentration found in rain.

Given that the half-life of tritium is 12·3 years, how old is the wine?

1

(3)

[Turn over

Marks

4. Hydrogen gas is widely regarded as a very valuable fuel for the future.

(a) Hydrogen can be produced from methane by steam reforming. The process proceeds in two steps.

Step 1: $CH_4(g)$ + $H_2O(g)$ → $CO(g)$ + $3H_2(g)$

Step 2: $CO(g)$ + $H_2O(g)$ → $CO_2(g)$ + $H_2(g)$

(i) What name is given to the gas mixture produced in step 1?

1

(ii) Using this process, how many moles of hydrogen gas can be produced overall from one mole of methane?

1

(b) Hydrogen can be produced in the lab from dilute sulphuric acid. The apparatus shown below can be used to investigate the quantity of electrical charge required to form one mole of hydrogen gas.

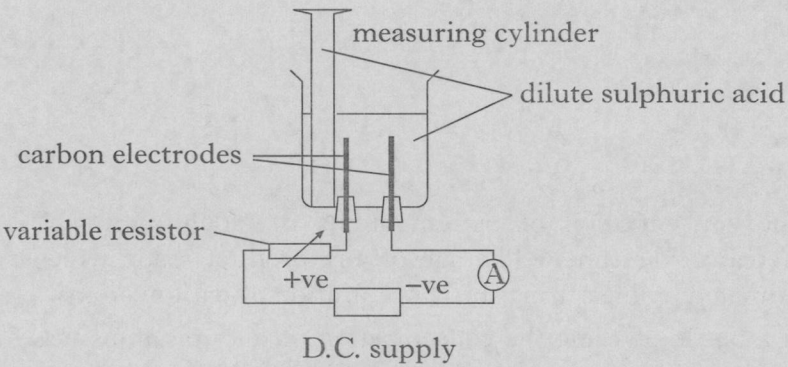

D.C. supply

(i) Above which electrode should the measuring cylinder be placed to collect the hydrogen gas?

1

(ii) In addition to the current, what **two** measurements should be taken?

1

(4)

DO NOT WRITE IN THIS MARGIN

Marks

5. The energy changes taking place during chemical reactions have many everyday uses.

(*a*) Some portable cold packs make use of the temperature drop that takes place when the chemicals in the pack dissolve in water.

Name the type of reaction that results in a fall in temperature. **1**

(*b*) Flameless heaters are used by mountain climbers to heat food and drinks. The chemical reaction in a flameless heater releases 45 kJ of energy.

If 200 g of water is heated using this heater, calculate the rise in temperature of the water, in °C. **1**

(2)

[Turn over

Marks

6. Temperature has a very significant effect on the rate of a chemical reaction.

(a) The reaction shown below can be used to investigate the effect of temperature on reaction rate.

$$5(COOH)_2(aq) + 6H^+(aq) + 2MnO_4^-(aq) \rightarrow 2Mn^{2+}(aq) + 10CO_2(g) + 8H_2O(\ell)$$

The instructions for such an investigation are shown below.

Procedure

1. Using syringes, add $5\,cm^3$ of sulphuric acid, $2\,cm^3$ of potassium permanganate solution and $40\,cm^3$ of water to a $100\,cm^3$ dry glass beaker.

2. Heat the mixture to about $40\,°C$.

3. Place the beaker on a white tile and measure $1\,cm^3$ of oxalic acid solution into a syringe.

4. Add the oxalic acid to the mixture in the beaker as quickly as possible and at the same time start the timer.

5. Gently stir the reaction mixture with the thermometer.

6. When the reaction is over, stop the timer and record the time. Measure and record the temperature of the reaction mixture.

7. Repeat the experiment three times but heat the initial sulphuric acid/potassium permanganate/water mixtures first to $50\,°C$, then to $60\,°C$ and finally to $70\,°C$.

(i) What colour change indicates that the reaction is over?

1

(ii) With each of the experiments, the temperature of the solution was measured both during heating and at the end of the reaction.

When plotting graphs of the reaction rate against temperature, it is the temperature measured at the end of reaction, rather than the temperature measured while heating, that is used.

Give a reason for this.

1

Marks

6. (continued)

(*b*) The graph shows the distribution of kinetic energy for molecules in a reaction mixture at a given temperature.

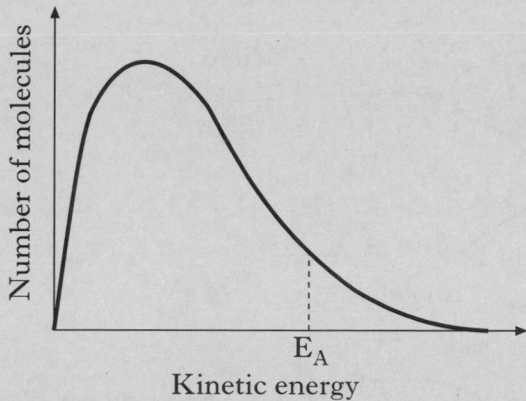

Why does a small increase in temperature produce a large increase in reaction rate?

1

(3)

[Turn over

Marks

7. Magnesium metal can be extracted from sea water.

An outline of the reactions involved is shown in the flow diagram.

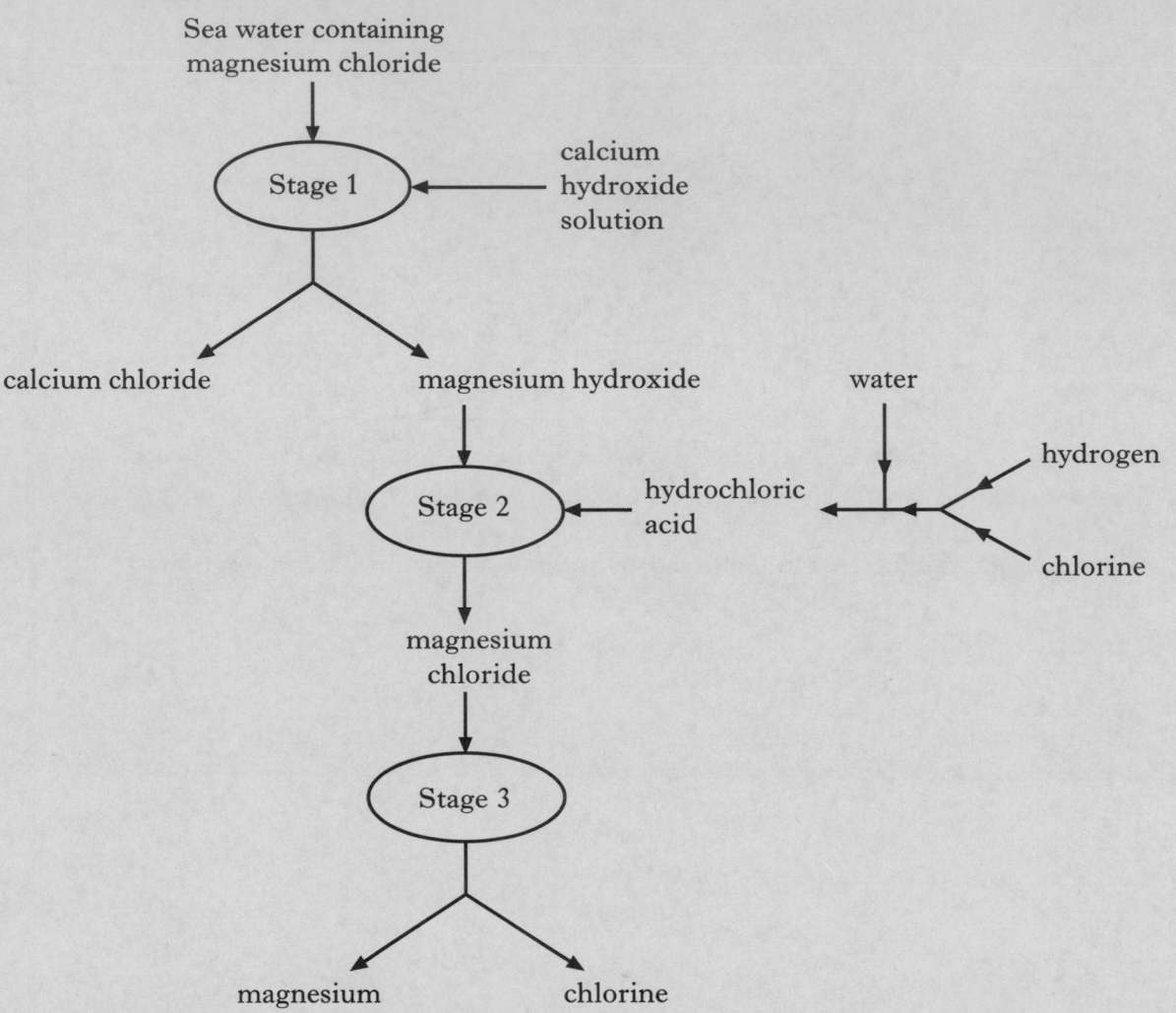

(a) Why can the magnesium hydroxide be easily separated from the calcium chloride at Stage 1?

1

(b) Name the type of chemical reaction taking place at Stage 2.

1

Marks

7. (continued)

(c) Give **two** different features of this process that make it economical.

2

(d) At Stage 3, electrolysis of molten magnesium chloride takes place.

If a current of 200 000 A is used, calculate the mass of magnesium, in kg, produced in 1 minute.

Show your working clearly.

3

(7)

[Turn over

Marks

8. One of the chemicals released in a bee sting is an ester that has the structure shown.

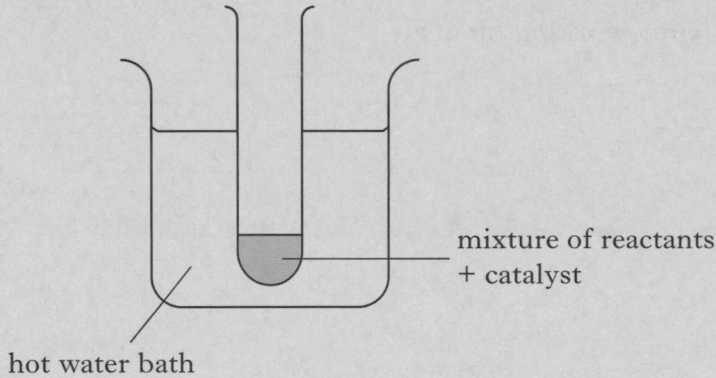

This ester can be produced by the reaction of an alcohol with an alkanoic acid.

(*a*) Name this acid.

1

(*b*) The ester can be prepared in the lab by heating a mixture of the reactants with a catalyst.

mixture of reactants + catalyst

hot water bath

(i) Name the catalyst used in the reaction.

1

(ii) What improvement could be made to the experimental set-up shown in the diagram?

1

Marks

8. (continued)

(*c*) If there is a 65% yield, calculate the mass of ester produced, in grams, when 4·0 g of the alcohol reacts with a slight excess of the acid.

(Mass of one mole of the alcohol = 88 g; mass of one mole of the ester = 130 g)

Show your working clearly.

2

(5)

[Turn over

Marks

9. Phenylalanine and alanine are both amino acids.

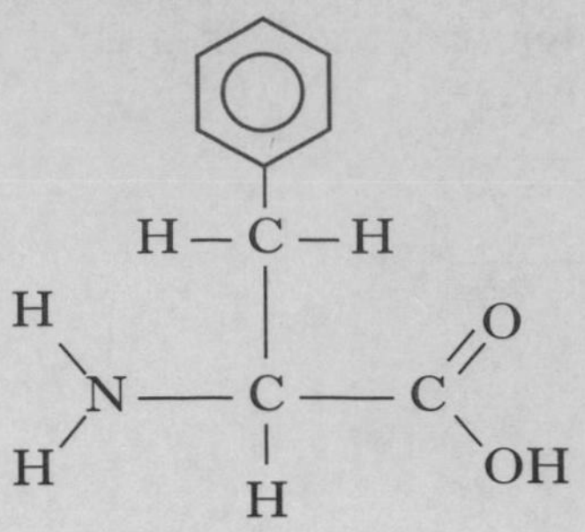

phenylalanine

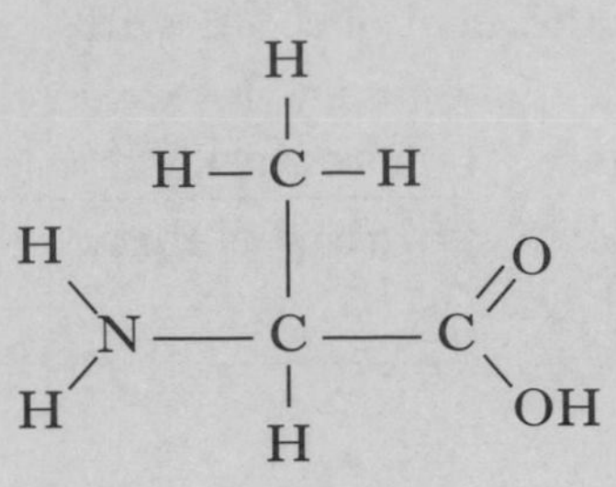

alanine

(a) Phenylalanine is an essential amino acid.

   (i) What is meant by an essential amino acid?

1

   (ii) How many hydrogen atoms are present in a molecule of phenylalanine?

1

(b) Phenylalanine and alanine can react to form the dipeptide shown.

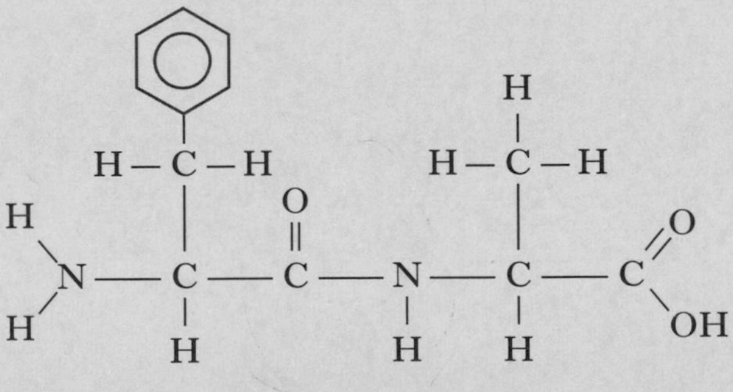

Circle the peptide link in this molecule.

1

(c) Draw a structural formula for the other dipeptide that can be formed from phenylalanine and alanine.

1

(4)

Marks

10. A student carried out three experiments involving the reaction of excess magnesium ribbon with dilute acids. The rate of hydrogen production was measured in each of the three experiments.

Experiment	Acid
1	$100\,cm^3$ of $0.10\,mol\,l^{-1}$ sulphuric acid
2	$50\,cm^3$ of $0.20\,mol\,l^{-1}$ sulphuric acid
3	$100\,cm^3$ of $0.10\,mol\,l^{-1}$ hydrochloric acid

The equation for **Experiment 1** is shown.

$$Mg(s) \quad + \quad H_2SO_4(aq) \quad \rightarrow \quad MgSO_4(aq) \quad + \quad H_2(g)$$

(*a*) The curve obtained for **Experiment 1** is drawn on the graph.

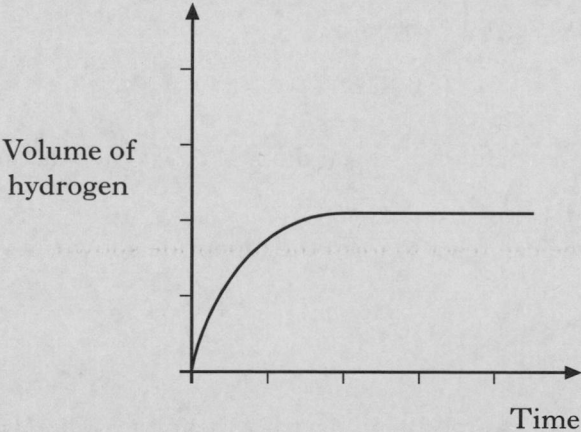

Draw curves on the graph to show the results obtained for **Experiment 2** and **Experiment 3**.

Label each curve clearly.

(An additional graph, if required, can be found on *Page thirty*.)

2

(*b*) The mass of magnesium used in **Experiment 1** was $0.50\,g$.

For this experiment, calculate the mass of magnesium, in grams, left unreacted.

2

(4)

Marks

11. Nitrogen dioxide gas can be prepared in different ways.

(a) It is manufactured industrially as part of the Ostwald process. In the first stage of the process, nitrogen monoxide is produced by passing ammonia and oxygen over a platinum catalyst.

$$NH_3(g) \quad + \quad O_2(g) \quad \rightarrow \quad NO(g) \quad + \quad H_2O(g)$$

   (i) Balance the above equation.

1

   (ii) Platinum metal is a heterogeneous catalyst for this reaction.

What is meant by a **heterogeneous** catalyst?

1

   (iii) The nitrogen monoxide then combines with oxygen in an exothermic reaction to form nitrogen dioxide.

$$2NO(g) \quad + \quad O_2(g) \quad \rightleftharpoons \quad 2NO_2(g)$$

What happens to the yield of nitrogen dioxide gas if the reaction mixture is cooled?

1

(b) In the lab, nitrogen dioxide gas can be prepared by heating copper(II) nitrate.

$$Cu(NO_3)_2(s) \quad \rightarrow \quad CuO(s) \quad + \quad 2NO_2(g) \quad + \quad \tfrac{1}{2}O_2(g)$$

   (i) Calculate the volume of nitrogen dioxide gas produced when $2 \cdot 0\,g$ of copper(II) nitrate is completely decomposed on heating.

(Take the molar volume of nitrogen dioxide to be 24 litres mol^{-1}.)

Show your working clearly.

2

Marks

11. (*b*) **(continued)**

(ii) Nitrogen dioxide has a boiling point of 22 °C.

Complete the diagram to show how nitrogen dioxide can be separated and collected.

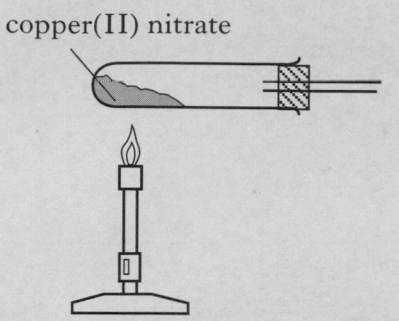

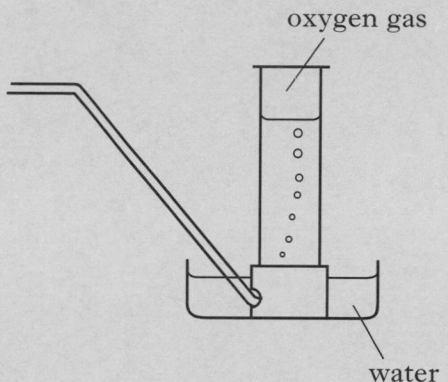

1

(6)

[Turn over

Marks

12. Ethyne is the first member of the homologous series called the alkynes.

(*a*) Ethyne can undergo addition reactions as shown in the flow diagram.

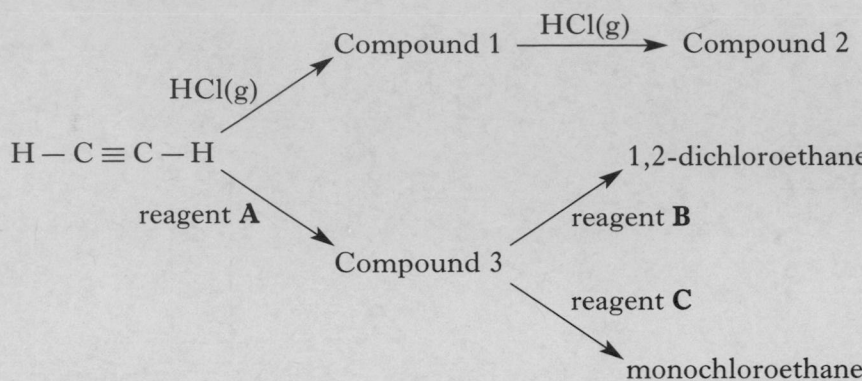

(i) Compound 2 is an isomer of 1,2-dichloroethane.

Draw a structural formula for compound 2.

1

(ii) Reagents **A**, **B** and **C** are three **different** diatomic gases.

Using information in the flow diagram, identify reagents **A**, **B** and **C**.

Reagent	Gas
A	
B	
C	

1

Marks

12. **(continued)**

(b) The equation for the enthalpy of formation of ethyne is:

$$2C(s) \quad + \quad H_2(g) \quad \rightarrow \quad C_2H_2(g)$$

Use the enthalpies of combustion of carbon, hydrogen and ethyne given in the data booklet to calculate the enthalpy of formation of ethyne, in $kJ\,mol^{-1}$.

Show your working clearly.

2

(4)

[Turn over

Marks

13. Compared to other gases made up of molecules of similar molecular masses, ammonia has a relatively high boiling point.

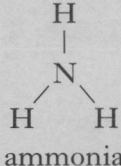

ammonia

(a) In terms of the intermolecular bonding present, **explain clearly** why ammonia has a relatively high boiling point.

2

(b) Amines can be produced by reacting ammonia with an aldehyde or a ketone. This reaction, an example of reductive amination, occurs in two stages.

Stage 1 Amination

$$CH_3 - \overset{\overset{O}{\|}}{C} - CH_3 \ + \ NH_3 \ \longrightarrow \ CH_3 - \overset{\overset{\overset{H}{|}}{N}}{\underset{\|}{C}} - CH_3$$

Stage 2 Reduction

$$CH_3 - \overset{\overset{\overset{H}{|}}{N}}{\underset{\|}{C}} - CH_3 \ \longrightarrow \ CH_3 - \overset{NH_2}{\underset{H}{C}} - CH_3$$

(i) Give another name for the type of reaction taking place in Stage 2.

1

Marks

13. (*b*) **(continued)**

(ii) Draw the structural formula for the amine produced when butanal undergoes reductive amination with ammonia.

1

(4)

[Turn over

Marks

14. Foodstuffs have labels that list ingredients and provide nutritional information.

    (*a*) The label on a tub of margarine lists **hydrogenated vegetable oils** as one of the ingredients.

       Why have some of the vegetable oils in this product been hydrogenated?

1

    (*b*) Potassium sorbate is a salt that is used as a preservative in margarine.

       Potassium sorbate dissolves in water to form an alkaline solution.

       What does this indicate about sorbic acid?

1

    (*c*) The nutritional information states that $100\,g$ of margarine contains $0.70\,g$ of sodium. The sodium is present as sodium chloride ($NaCl$).

       Calculate the mass of sodium chloride, in g, present in every $100\,g$ of margarine.

1

(3)

Marks

15. Seaweeds are a rich source of iodine in the form of iodide ions. The mass of iodine in a seaweed can be found using the procedure outlined below.

(a) **Step 1**

The seaweed is dried in an oven and ground into a fine powder. Hydrogen peroxide solution is then added to oxidise the iodide ions to iodine molecules. The ion-electron equation for the reduction reaction is shown.

$$H_2O_2(aq) \quad + \quad 2H^+(aq) \quad + \quad 2e^- \quad \rightarrow \quad 2H_2O(\ell)$$

Write a balanced redox equation for the reaction of hydrogen peroxide with iodide ions.

1

(b) **Step 2**

Using starch solution as an indicator, the iodine solution is then titrated with sodium thiosulphate solution to find the mass of iodine in the sample. The balanced equation for the reaction is shown.

$$2Na_2S_2O_3(aq) \quad + \quad I_2(aq) \quad \rightarrow \quad 2NaI(aq) \quad + \quad Na_2S_4O_6(aq)$$

In an analysis of seaweed, $14 \cdot 9 \, cm^3$ of $0 \cdot 00500 \, mol\,l^{-1}$ sodium thiosulphate solution was required to reach the end-point.

Calculate the mass of iodine present in the seaweed sample.

Show your working clearly.

3

(4)

[END OF QUESTION PAPER]

SPACE FOR ANSWERS

ADDITIONAL GRAPH FOR QUESTION 10(*a*)

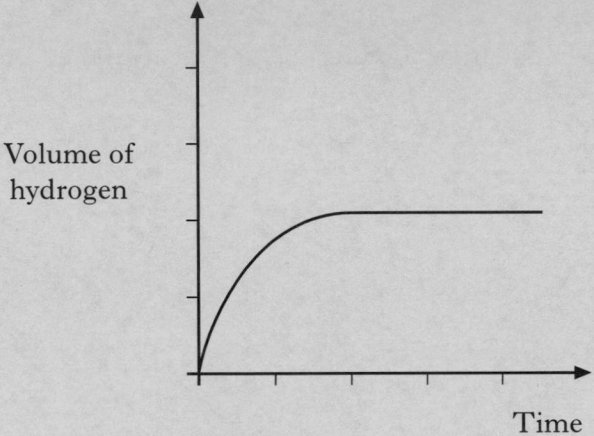

ADDITIONAL SPACE FOR ANSWERS

ADDITIONAL SPACE FOR ANSWERS

[BLANK PAGE]

[BLANK PAGE]

FOR OFFICIAL USE

Total
Section B

X012/301

NATIONAL
QUALIFICATIONS
2008

FRIDAY, 30 MAY
9.00 AM – 11.30 AM

CHEMISTRY
HIGHER

Fill in these boxes and read what is printed below.

Full name of centre

Town

Forename(s)

Surname

Date of birth
Day Month Year

Scottish candidate number

Number of seat

Reference may be made to the Chemistry Higher and Advanced Higher Data Booklet.

SECTION A—Questions 1–40 (40 marks)

Instructions for completion of **Section A** are given on page two.

For this section of the examination you must use an **HB pencil**.

SECTION B (60 marks)

1   All questions should be attempted.

2   The questions may be answered in any order but all answers are to be written in the spaces provided in this answer book, **and must be written clearly and legibly in ink**.

3   Rough work, if any should be necessary, should be written in this book and then scored through when the fair copy has been written. If further space is required, a supplementary sheet for rough work may be obtained from the invigilator.

4   Additional space for answers will be found at the end of the book. If further space is required, supplementary sheets may be obtained from the invigilator and should be inserted inside the **front** cover of this book.

5   The size of the space provided for an answer should not be taken as an indication of how much to write. It is not necessary to use all the space.

6   Before leaving the examination room you must give this book to the invigilator. If you do not, you may lose all the marks for this paper.

SECTION A

Read carefully

1   Check that the answer sheet provided is for **Chemistry Higher (Section A)**.

2   For this section of the examination you must use an **HB pencil** and, where necessary, an eraser.

3   Check that the answer sheet you have been given has **your name**, **date of birth**, **SCN** (Scottish Candidate Number) and **Centre Name** printed on it.

   Do not change any of these details.

4   If any of this information is wrong, tell the Invigilator immediately.

5   If this information is correct, **print** your name and seat number in the boxes provided.

6   The answer to each question is **either** A, B, C or D. Decide what your answer is, then, using your pencil, put a horizontal line in the space provided (see sample question below).

7   There is **only one correct** answer to each question.

8   Any rough working should be done on the question paper or the rough working sheet, **not** on your answer sheet.

9   At the end of the exam, put the **answer sheet for Section A inside the front cover of your answer book**.

Sample Question

To show that the ink in a ball-pen consists of a mixture of dyes, the method of separation would be

   A   chromatography

   B   fractional distillation

   C   fractional crystallisation

   D   filtration.

The correct answer is **A**—chromatography. The answer **A** has been clearly marked in **pencil** with a horizontal line (see below).

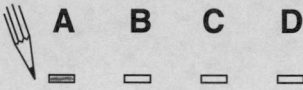

Changing an answer

If you decide to change your answer, carefully erase your first answer and using your pencil, fill in the answer you want. The answer below has been changed to **D**.

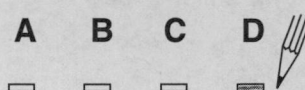

1. Solutions of barium chloride and silver nitrate are mixed together.

   The reaction that takes place is an example of

   A displacement

   B neutralisation

   C oxidation

   D precipitation.

2. Two rods are placed in dilute sulphuric acid as shown.

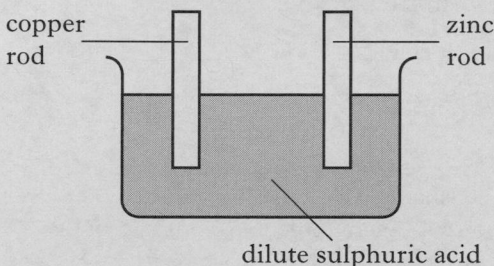

   copper rod

   zinc rod

   dilute sulphuric acid

   Which of the following would be observed?

   A No gas is given off.

   B Gas is given off at only the zinc rod.

   C Gas is given off at only the copper rod.

   D Gas is given off at both rods.

3. An element was burned in air. The product was added to water, producing a solution with a pH less than 7.

   The element could be

   A carbon

   B hydrogen

   C sodium

   D tin.

4. A mixture of sodium chloride and sodium sulphate is known to contain 0·6 mol of chloride ions and 0·2 mol of sulphate ions.

   How many moles of sodium ions are present?

   A 0·4

   B 0·5

   C 0·8

   D 1·0

5. The following results were obtained in the reaction between marble chips and dilute hydrochloric acid.

Time/minutes	0	2	4	6	8	10
Total volume of carbon dioxide produced/cm^3	0	52	68	78	82	84

   What is the average rate of production of carbon dioxide, in $cm^3\ min^{-1}$, between 2 and 8 minutes?

   A 5

   B 26

   C 30

   D 41

6. 5 g of copper is added to excess silver nitrate solution. The equation for the reaction that takes place is:

   $$Cu(s) + 2AgNO_3(aq) \rightarrow 2Ag(s) + Cu(NO_3)_2(aq)$$

   After some time, the solid present is filtered off from the solution, washed with water, dried and weighed.

   The final mass of the solid will be

   A less than 5 g

   B 5 g

   C 10 g

   D more than 10 g.

7.

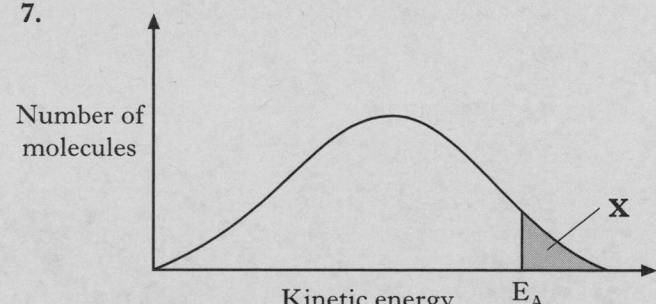

   Number of molecules

   Kinetic energy E_A

   In area **X**

   A molecules always form an activated complex

   B no molecules have the energy to form an activated complex

   C collisions between molecules are always successful in forming products

   D all molecules have the energy to form an activated complex.

8.

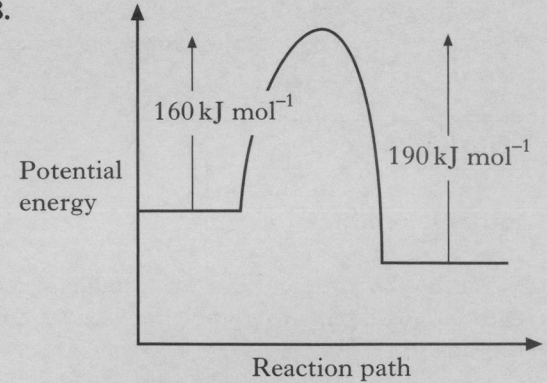

When a catalyst is used, the activation energy of the forward reaction is reduced to $35\,kJ\,mol^{-1}$.

What is the activation energy of the catalysed reverse reaction?

A $30\,kJ\,mol^{-1}$

B $35\,kJ\,mol^{-1}$

C $65\,kJ\,mol^{-1}$

D $190\,kJ\,mol^{-1}$

9. As the atomic number of the alkali metals increases

A   the first ionisation energy decreases

B   the atomic size decreases

C   the density decreases

D   the melting point increases.

10. Which of the following atoms has the least attraction for bonding electrons?

A   Carbon

B   Nitrogen

C   Phosphorus

D   Silicon

11. Which of the following reactions refers to the third ionisation energy of aluminium?

A $Al(s) \rightarrow Al^{3+}(g) + 3e^-$

B $Al(g) \rightarrow Al^{3+}(g) + 3e^-$

C $Al^{2+}(g) \rightarrow Al^{3+}(g) + e^-$

D $Al^{3+}(g) \rightarrow Al^{4+}(g) + e^-$

12. Which of the following represents an exothermic process?

A $Cl_2(g) \rightarrow 2Cl(g)$

B $Na(s) \rightarrow Na(g)$

C $Na(g) \rightarrow Na^+(g) + e^-$

D $Na^+(g) + Cl^-(g) \rightarrow Na^+Cl^-(s)$

13. In which of the following liquids does hydrogen bonding occur?

A   Ethanol

B   Ethyl ethanoate

C   Hexane

D   Pent-1-ene

14. The shapes of some common molecules are shown. Each molecule contains at least one polar covalent bond.

Which of the following molecules is non-polar?

A $H - Cl$

B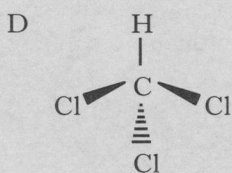

C $O = C = O$

D
$$\begin{array}{c} H \\ | \\ Cl \underset{\vdots}{\overset{C}{\diagup \diagdown}} Cl \\ Cl \end{array}$$

15. At room temperature, a solid substance was shown to have a lattice consisting of positively charged ions and delocalised outer electrons.

The substance could be

A   graphite

B   sodium

C   mercury

D   phosphorus.

16. The mass of 1 mol of sodium is 23 g.

    What is the approximate mass of one sodium atom?

    A 6×10^{23} g

    B 6×10^{-23} g

    C $3 \cdot 8 \times 10^{-23}$ g

    D $3 \cdot 8 \times 10^{-24}$ g

17. In which of the following pairs do the gases contain the same number of oxygen atoms?

    A 1 mol of oxygen and 1 mol of carbon monoxide

    B 1 mol of oxygen and 0·5 mol of carbon dioxide

    C 0·5 mol of oxygen and 1 mol of carbon dioxide

    D 1 mol of oxygen and 1 mol of carbon dioxide

18. The Avogadro Constant is the same as the number of

    A molecules in 16 g of oxygen

    B electrons in 1 g of hydrogen

    C atoms in 24 g of carbon

    D ions in 1 litre of sodium chloride solution, concentration $1 \, mol \, l^{-1}$.

19. $$2NO(g) + O_2(g) \rightarrow 2NO_2(g)$$

    How many litres of nitrogen dioxide gas would be produced in a reaction, starting with a mixture of 5 litres of nitrogen monoxide gas and 2 litres of oxygen gas?

    (All volumes are measured under the same conditions of temperature and pressure.)

    A 2

    B 3

    C 4

    D 5

20. Which of the following fuels can be produced by the fermentation of biological material under anaerobic conditions?

    A Hydrogen

    B Methane

    C Methanol

    D Petrol

21. Butadiene is the first member of a homologous series of hydrocarbons called dienes.

    What is the general formula for this series?

    A C_nH_{n+2}

    B C_nH_{n+3}

    C C_nH_{2n}

    D C_nH_{2n-2}

22.

$$CH_3 - CH = CH_2$$

Reaction **X**

$$CH_3 - CH_2 - CH_2 - OH$$

Reaction **Y**

$$CH_3 - CH_2 - C \overset{O}{\underset{H}{}}$$

Which line in the table correctly describes reactions **X** and **Y**?

	Reaction X	**Reaction Y**
A	hydration	oxidation
B	hydration	reduction
C	hydrolysis	oxidation
D	hydrolysis	reduction

23. Ammonia is manufactured from hydrogen and nitrogen by the Haber Process.

    $$3H_2(g) \quad + \quad N_2(g) \rightleftharpoons 2NH_3(g)$$

    If 80 kg of ammonia is produced from 60 kg of hydrogen, what is the percentage yield?

    A $\dfrac{80}{340} \times 100$

    B $\dfrac{80}{170} \times 100$

    C $\dfrac{30}{80} \times 100$

    D $\dfrac{60}{80} \times 100$

    [Turn over

24. Which of the following statements about methanol is **false**?

A It can be made from synthesis gas.

B It can be dehydrated to form an alkene.

C It can be oxidised to give a carboxylic acid.

D It reacts with acidified potassium dichromate solution.

25. A by-product produced in the manufacture of a polyester has the structure shown.

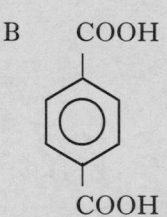

What is the structure of the diacid monomer used in the polymerisation?

A   COOH

         COOH

B   COOH

         COOH

C HOOC — CH₂ — CH₂ — COOH

D   COOH

         CH₂CH₂COOH

26. Which of the following statements can be applied to polymeric esters?

A They are used for flavourings, perfumes and solvents.

B They are manufactured for use as textile fibres and resins.

C They are cross-linked addition polymers.

D They are condensation polymers made by the linking up of amino acids.

27. The rate of hydrolysis of protein, using an enzyme, was studied at different temperatures.

Which of the following graphs would be obtained?

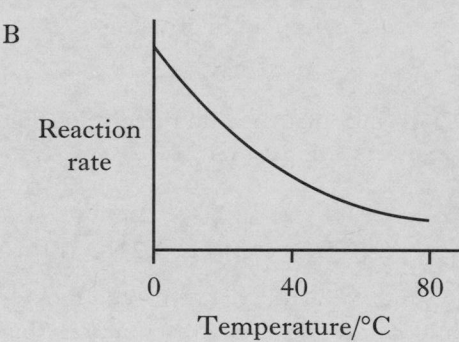

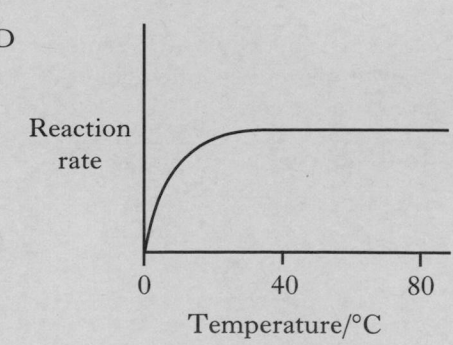

28. Which of the following arrangements of atoms shows a peptide link?

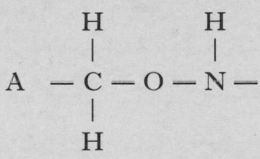

```
        H           H
        |           |
A    — C — O — N —
        |
        H

        H   O   H
        |   ||  |
B    — C — C — N —
        |
        H

        H   OH
        |   |
C    — C — C = N —
        |
        H

        H   O       H
        |   ||      |
D    — C — C — O — N —
        |
        H
```

29. Which line in the table shows the effect of a catalyst on the reaction rates and position of equilibrium in a reversible reaction?

	Rate of forward reaction	Rate of reverse reaction	Position of equilibrium
A	increased	unchanged	moves right
B	increased	increased	unchanged
C	increased	decreased	moves right
D	unchanged	unchanged	unchanged

30. The following equilibrium exists in bromine water.

$$Br_2(aq) + H_2O(\ell) \rightleftharpoons Br^-(aq) + 2H^+(aq) + OBr^-(aq)$$
(red)   (colourless)   (colourless)

The red colour of bromine water would fade on adding a few drops of a concentrated solution of

A HCl

B KBr

C $AgNO_3$

D NaOBr.

31. Which of the following is the best description of a $0 \cdot 1$ mol l^{-1} solution of nitric acid?

A Dilute solution of a weak acid

B Dilute solution of a strong acid

C Concentrated solution of a weak acid

D Concentrated solution of a strong acid

32. The conductivity of pure water is low because

A water molecules are polar

B only a few water molecules are ionised

C water molecules are linked by hydrogen bonds

D there are equal numbers of hydrogen and hydroxide ions in water.

33. Which of the following statements is **true** about an aqueous solution of ammonia?

A It has a pH less than 7.

B It is completely ionised.

C It contains more hydroxide ions than hydrogen ions.

D It reacts with acids producing ammonia gas.

34. Equal volumes of solutions of ethanoic acid and hydrochloric acid, of equal concentration, are compared.

In which of the following cases does the ethanoic acid give the higher value?

A pH of solution

B Conductivity of solution

C Rate of reaction with magnesium

D Volume of sodium hydroxide solution neutralised

35. Equal volumes of $0 \cdot 1$ mol l^{-1} solutions of the following acids and alkalis were mixed.

Which of the following pairs would give the solution with the lowest pH?

A Hydrochloric acid and sodium hydroxide

B Hydrochloric acid and calcium hydroxide

C Sulphuric acid and sodium hydroxide

D Sulphuric acid and calcium hydroxide

[Turn over

36. Which of the following compounds dissolves in water to form an acidic solution?

   A   Sodium nitrate

   B   Barium sulphate

   C   Potassium ethanoate

   D   Ammonium chloride

37. The ion-electron equations for a redox reaction are:

$$2I^-(aq) \rightarrow I_2(aq) + 2e^-$$

$$MnO_4^-(aq) + 8H^+(aq) + 5e^- \rightarrow Mn^{2+}(aq) + 4H_2O(\ell)$$

How many moles of iodide ions are oxidised by one mole of permanganate ions?

   A   0·2

   B   0·4

   C   2

   D   5

38. In which of the following reactions is the hydrogen ion acting as an oxidising agent?

   A $Mg + 2HCl \rightarrow MgCl_2 + H_2$

   B $NaOH + HNO_3 \rightarrow NaNO_3 + H_2O$

   C $CuCO_3 + H_2SO_4 \rightarrow CuSO_4 + H_2O + CO_2$

   D $CH_3COONa + HCl \rightarrow NaCl + CH_3COOH$

39. An atom of ^{227}Th decays by a series of alpha emissions to form an atom of ^{211}Pb.

How many alpha particles are released in the process?

   A   2

   B   3

   C   4

   D   5

40. The half-life of the isotope ^{210}Pb is 21 years.

What fraction of the original ^{210}Pb atoms will be present after 63 years?

   A   0·5

   B   0·25

   C   0·125

   D   0·0625

Candidates are reminded that the answer sheet MUST be returned INSIDE the front cover of this answer book.

Marks

SECTION B

All answers must be written clearly and legibly in ink.

1. The formulae for three oxides of sodium, carbon and silicon are Na_2O, CO_2 and SiO_2.

   Complete the table for CO_2 and SiO_2 to show both the bonding and structure of the three oxides at room temperature.

Oxide	Bonding and structure
Na_2O	ionic lattice
CO_2	
SiO_2	

   (2)

2. A typical triglyceride found in olive oil is shown below.

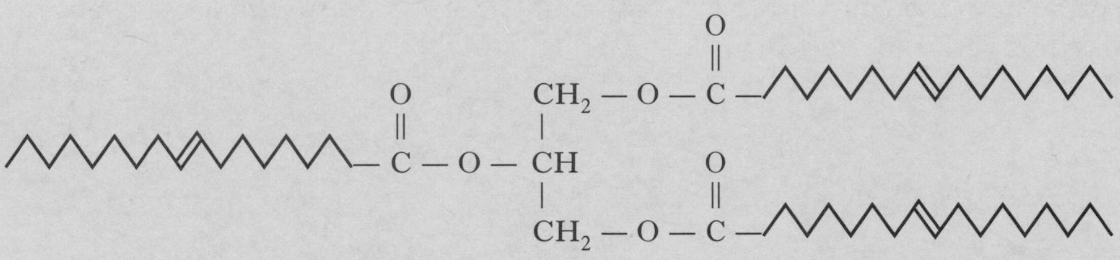

   (a) To which family of organic compounds do triglycerides belong?

   1

   (b) Olive oil can be hardened for use in margarines.

   What happens to the triglyceride molecules during the hardening of olive oil?

   1

   (c) Give **one** reason why oils can be a useful part of a balanced diet.

   1
   (3)

Page nine **[Turn over**

Marks

3. A student carried out the Prescribed Practical Activity (PPA) to find the effect of concentration on the rate of the reaction between hydrogen peroxide solution and an acidified solution of iodide ions.

$$H_2O_2(aq) \quad + \quad 2H^+(aq) \quad + \quad 2I^-(aq) \quad \rightarrow \quad 2H_2O(\ell) \quad + \quad I_2(aq)$$

During the investigation, only the concentration of the iodide ions was changed.

Part of the student's results sheet for this PPA is shown.

Results

Experiment	Volume of KI(aq) /cm^3	Volume of H$_2$O /cm^3	Volume of H$_2$O$_2$(aq) /cm^3	Volume of H$_2$SO$_4$(aq) /cm^3	Volume of Na$_2$S$_2$O$_3$(aq) /cm^3	Rate /s^{-1}
1	25	0	5	10	10	0·043
2						
3						

(a) Describe how the concentration of the potassium iodide solution was changed during this series of experiments.

1

(b) Calculate the reaction time, in seconds, for the first experiment.

1

(2)

Marks

4. Using a cobalt catalyst, alkenes react with a mixture of hydrogen and carbon monoxide.

The products are two isomeric aldehydes.

Propene reacts with the mixture as shown.

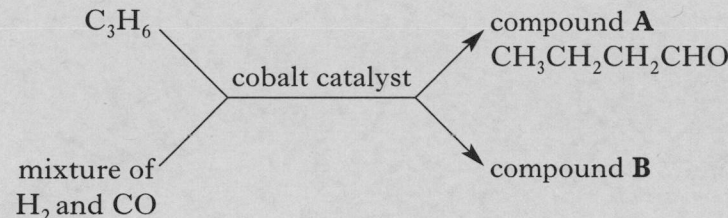

C$_3$H$_6$ → compound **A**
CH$_3$CH$_2$CH$_2$CHO

cobalt catalyst → compound **B**

mixture of
H$_2$ and CO

(*a*) What name is given to a mixture of hydrogen and carbon monoxide?

1

(*b*) Draw a structural formula for compound **B**.

1

(*c*)   (i) What would be observed if compound **A** was gently heated with Tollens' reagent?

1

  (ii) How would the reaction mixture be heated?

1

(*d*) Aldehydes can also be formed by the reaction of some alcohols with copper(II) oxide.

Name the **type** of alcohol that would react with copper(II) oxide to form an aldehyde.

1

(5)

[Turn over

Marks

5. All the isotopes of technetium are radioactive.

(*a*) Technetium-99 is produced as shown.

$$^{99}_{42}\text{Mo} \rightarrow {}^{99}_{43}\text{Tc} + \mathbf{X}$$

Identify **X**.

1

(*b*) The graph shows the decay curve for a 1·0 g sample of technetium-99.

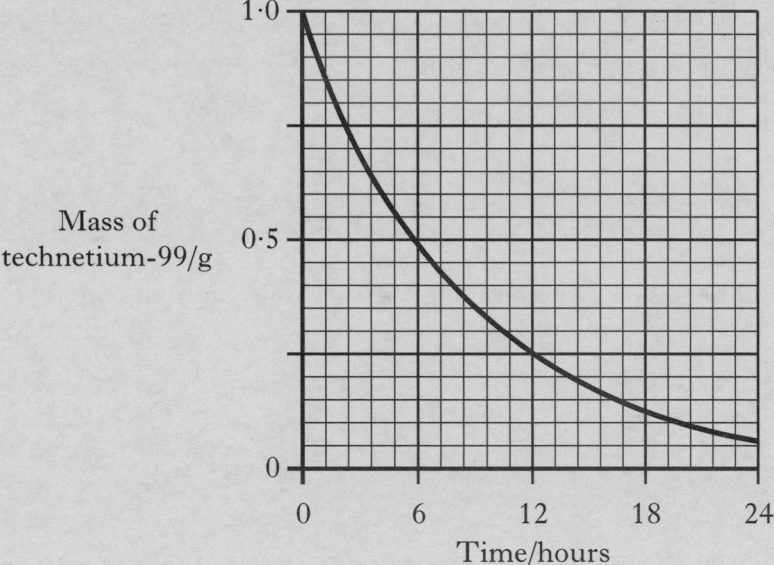

(i) Draw a curve on the graph to show the variation of mass with time for a 0·5 g sample of technetium-99.

(An additional graph, if required, can be found on *Page twenty-eight*.)

1

(ii) Technetium-99 is widely used in medicine to detect damage to heart tissue. It is a gamma-emitting radioisotope and is injected into the body.

Suggest **one** reason why technetium-99 can be safely used in this way.

1

(3)

6. In 1865, the German chemist Kekulé proposed a ring structure for benzene. This structure was based on alternating single and double bonds.

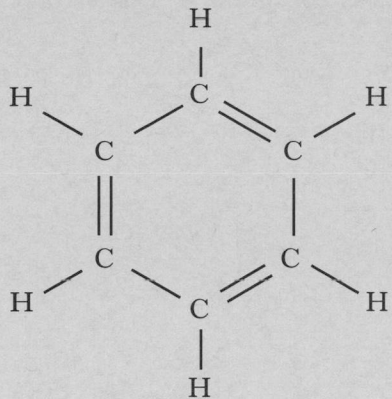

Marks

(a) (i) Describe a chemical test that would indicate that the above chemical structure for benzene is incorrect.

1

    (ii) Briefly describe the correct structure for benzene.

1

(b) Benzene can be formed from cyclohexane.

$$C_6H_{12} \rightarrow C_6H_6 + 3H_2$$

What name is given to this type of reaction?

1

(c) Benzene is also added in very small amounts to some petrols. Why is benzene added to petrol?

1

(4)

Marks

7. Hydrogen fluoride, HF, is used to manufacture hydrofluorocarbons.

   Hydrofluorocarbons are now used as refrigerants instead of chlorofluorocarbons, CFCs.

   (a) Why are CFCs no longer used?

   1

   (b) Hydrogen fluoride gas is manufactured by reacting calcium fluoride with concentrated sulphuric acid.

   $$CaF_2 \ + \ H_2SO_4 \ \rightarrow \ CaSO_4 \ + \ 2HF$$

   What volume of hydrogen fluoride gas is produced when $1{\cdot}0\,kg$ of calcium fluoride reacts completely with concentrated sulphuric acid?

   (Take the molar volume of hydrogen fluoride gas to be 24 litres mol^{-1}.)

   Show your working clearly.

   2

   (3)

Marks

8. Carbon monoxide can be produced in many ways.

(a) One method involves the reaction of carbon with an oxide of boron.

$$B_2O_3 \quad + \quad C \quad \rightarrow \quad B_4C \quad + \quad CO$$

Balance this equation.

1

(b) Carbon monoxide is also a product of the reaction of carbon dioxide with hot carbon. The carbon dioxide is made by the reaction of dilute hydrochloric acid with solid calcium carbonate.

Unreacted carbon dioxide is removed before the carbon monoxide is collected by displacement of water.

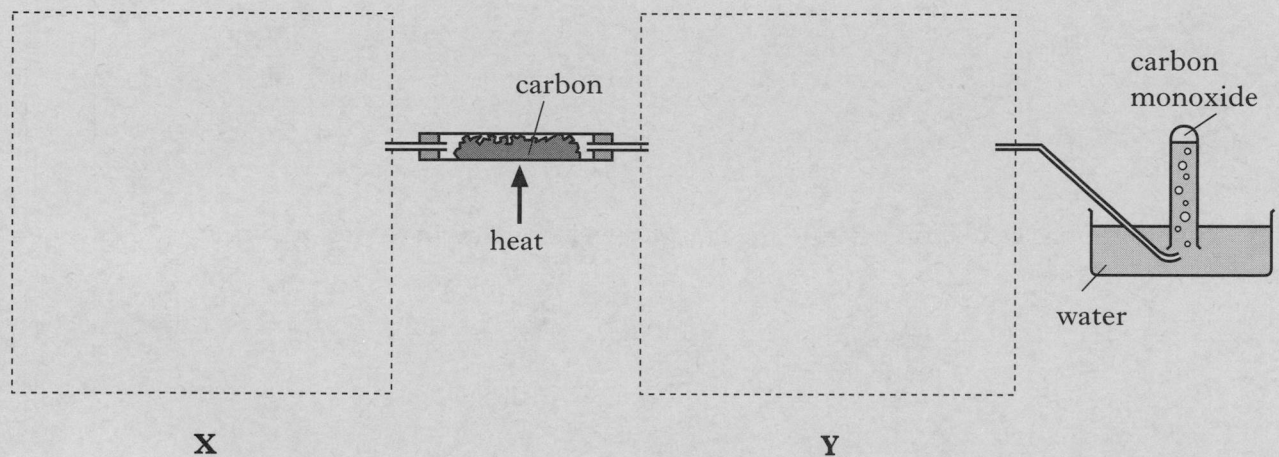

X   **Y**

Complete the diagram to show how the carbon dioxide can be produced at **X** and how the unreacted carbon dioxide can be removed by bubbling it through a solution at **Y**.

Normal laboratory apparatus should be used in your answer and the chemicals used at **X** and **Y** should be labelled.

2

(c) Why is carbon monoxide present in car exhaust fumes?

1

(4)

[Turn over

Marks

9. Hydrogen gas has a boiling point of −253 °C.

   (a) Explain clearly why hydrogen is a gas at room temperature.

   In your answer you should name the intermolecular forces involved and indicate how they arise.

   2

   (b) Hydrogen gas can be prepared in the lab by the electrolysis of dilute sulphuric acid.

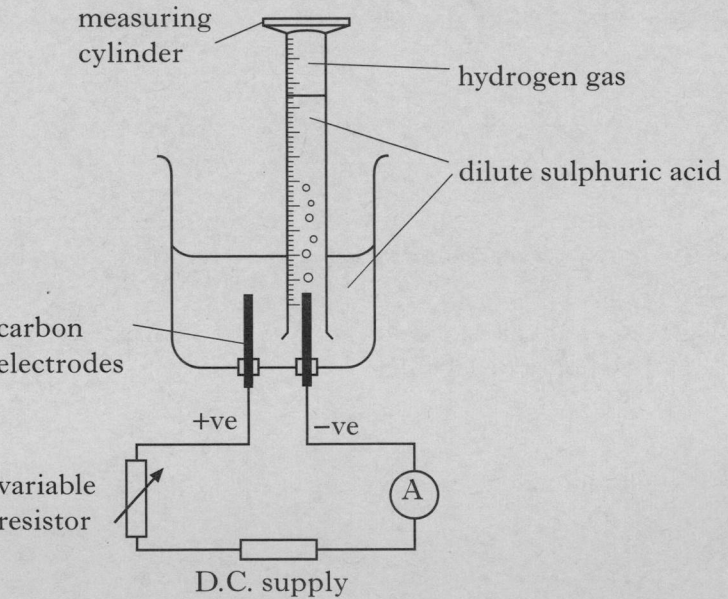

   (i) Before collecting the gas in the measuring cylinder, it is usual to switch on the current and allow bubbles of gas to be produced for a few minutes.

   Why is this done?

   1

Marks

9. **(b)** **(continued)**

(ii) The equation for the reaction at the negative electrode is:

$$2H^+(aq) \quad + \quad 2e^- \quad \rightarrow \quad H_2(g)$$

Calculate the mass of hydrogen gas, in grams, produced in 10 minutes when a current of $0\cdot30$ A was used.

Show your working clearly.

2

(c) The concentration of $H^+(aq)$ ions in the dilute sulphuric acid used in the experiment was 1×10^{-1} mol l^{-1}.

Calculate the concentration of $OH^-(aq)$ ions, in mol l^{-1}, in the dilute sulphuric acid.

1

(6)

[Turn over

Marks

10. When cyclopropane gas is heated over a catalyst, it isomerises to form propene gas and an equilibrium is obtained.

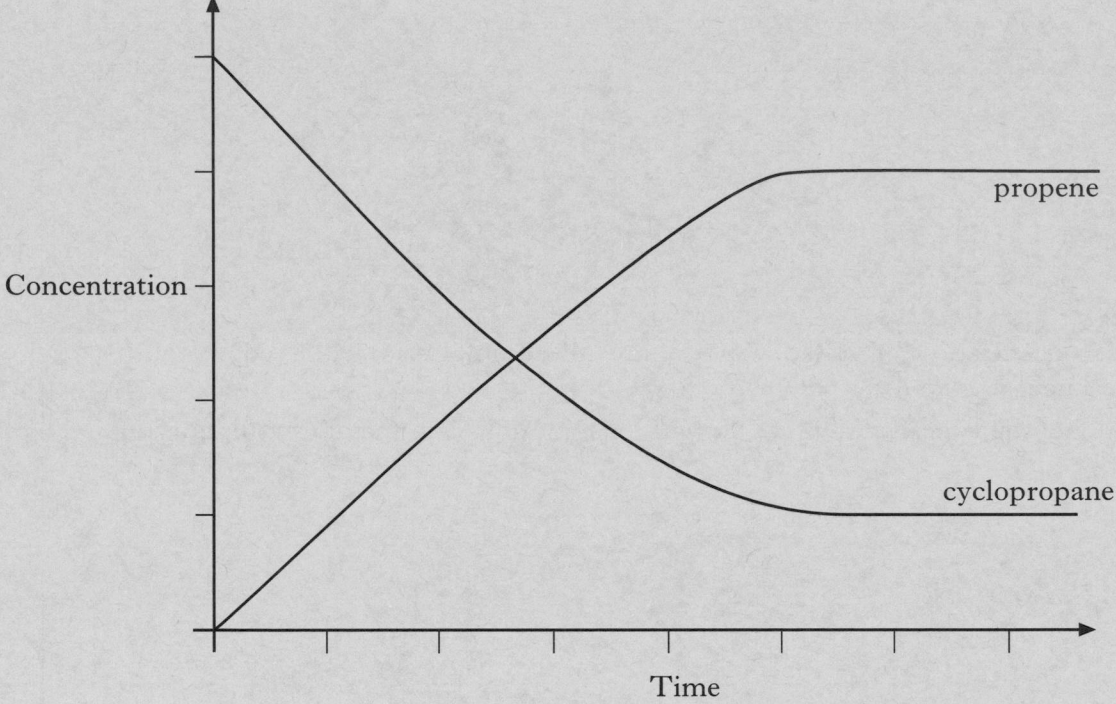

The graph shows the concentrations of cyclopropane and propene as equilibrium is established in the reaction.

(*a*) Mark clearly on the graph the point at which equilibrium has just been reached.

1

(*b*) Why does increasing the pressure have **no** effect on the position of this equilibrium?

1

Marks

10. **(continued)**

(*c*) The equilibrium can also be achieved by starting with propene.

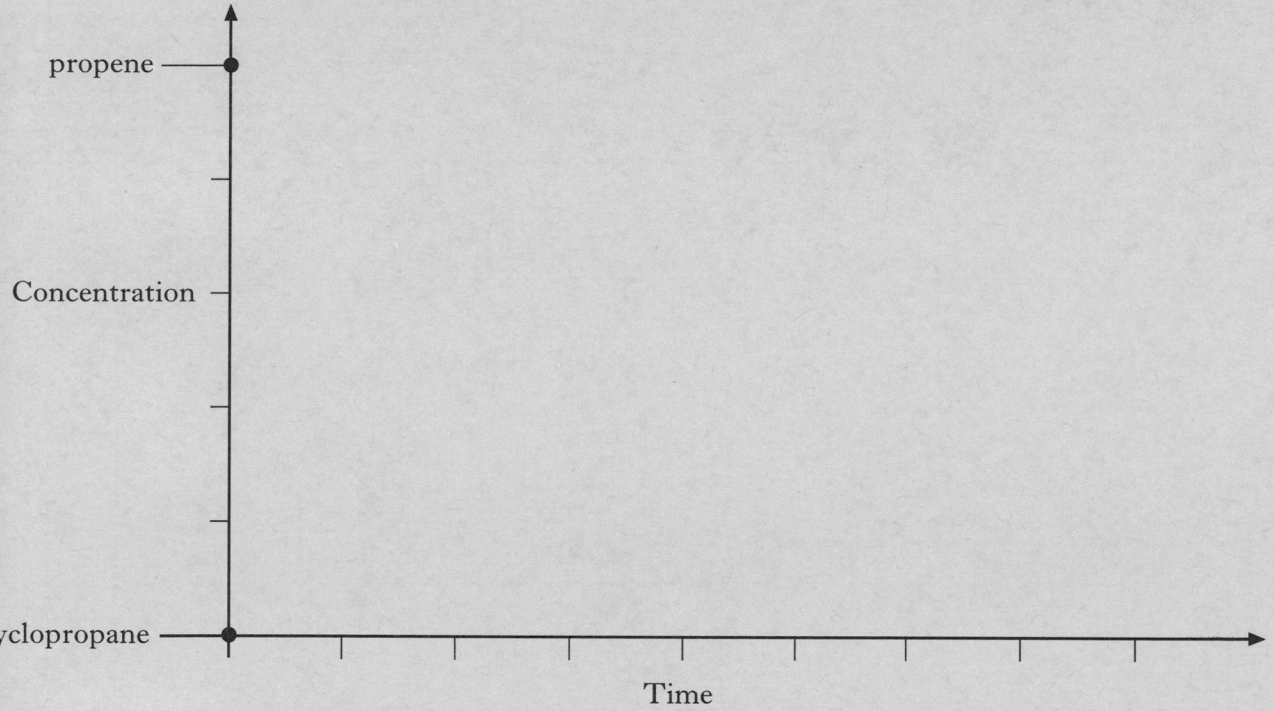

propene   cyclopropane

Using the initial concentrations shown, sketch a graph to show how the concentrations of propene and cyclopropane change as equilibrium is reached for this reverse reaction.

1

(3)

[Turn over

Marks

11. A student writes the following two statements. **Both are incorrect**.
In each case explain the mistake in the student's reasoning.

(*a*) Alcohols are alkaline because of their OH groups.

1

(*b*) Because of the iodine, potassium iodide will produce a blue/black colour in contact with starch.

1
(2)

DO NOT WRITE IN THIS MARGIN

Marks

12. When in danger, bombardier beetles can fire a hot, toxic mixture of chemicals at the attacker.

This mixture contains quinone, $C_6H_4O_2$, a compound that is formed by the reaction of hydroquinone, $C_6H_4(OH)_2$, with hydrogen peroxide, H_2O_2. The reaction is catalysed by an enzyme called catalase.

(*a*) Most enzymes can catalyse only specific reactions, eg catalase cannot catalyse the hydrolysis of starch.

Give a reason for this. **1**

(*b*) The equation for the overall reaction is:

$$C_6H_4(OH)_2(aq) \;+\; H_2O_2(aq) \;\rightarrow\; C_6H_4O_2(aq) \;+\; 2H_2O(\ell)$$

Use the following data to calculate the enthalpy change, in kJ mol^{-1}, for the above reaction.

$$C_6H_4(OH)_2(aq) \rightarrow C_6H_4O_2(aq) + H_2(g) \qquad \Delta H = +177{\cdot}4\,kJ\,mol^{-1}$$

$$H_2(g) + O_2(g) \rightarrow H_2O_2(aq) \qquad \Delta H = -191{\cdot}2\,kJ\,mol^{-1}$$

$$H_2(g) + \tfrac{1}{2}O_2(g) \rightarrow H_2O(g) \qquad \Delta H = -241{\cdot}8\,kJ\,mol^{-1}$$

$$H_2O(g) \rightarrow H_2O(\ell) \qquad \Delta H = -43{\cdot}8\,kJ\,mol^{-1}$$

Show your working clearly. **2**

(3)

[Turn over

Marks

13. For many years, carbohydrates found in plants have been used to provide chemicals. Lactic acid can be produced by fermenting the carbohydrates in corn.

Lactic acid has the structure:

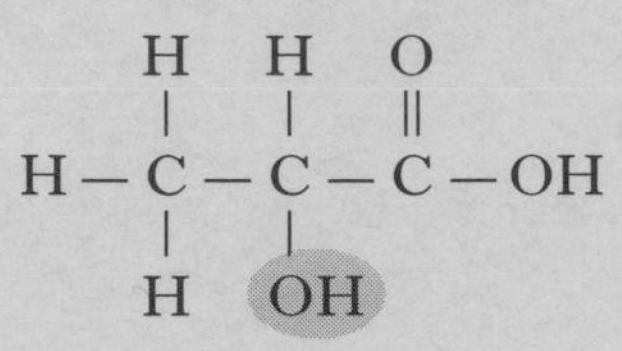

$$
\begin{array}{ccc}
H & H & O \\
| & | & \| \\
H-C-C-C-OH \\
| & | \\
H & OH
\end{array}
$$

(a) Name the functional group in the shaded area.

1

(b) Lactic acid is used to make polylactic acid, a biodegradeable polymer that is widely used for food packaging.

(i) Name another biodegradeable polymer.

1

(ii) Polylactic acid can be manufactured by either a batch or a continuous process.

What is meant by a batch process?

1

(iii) The first stage in the polymerisation of lactic acid involves the condensation of two lactic acid molecules to form a cyclic structure called a lactone.

Draw a structural formula for the lactone formed when two molecules of lactic acid undergo condensation with each other.

1

(4)

Marks

14. Hydrogen peroxide decomposes as shown:

$$H_2O_2(aq) \rightarrow H_2O(\ell) + \tfrac{1}{2}O_2(g)$$

The reaction can be catalysed by iron(III) nitrate solution.

(a) What **type** of catalyst is iron(III) nitrate solution in this reaction?

1

(b) In order to calculate the enthalpy change for the decomposition of hydrogen peroxide, a student added iron(III) nitrate solution to hydrogen peroxide solution.

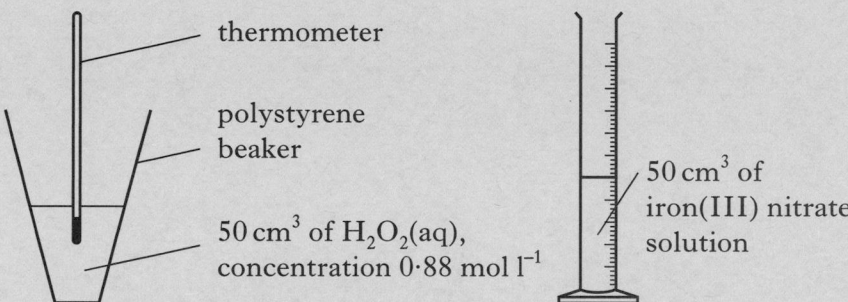

thermometer

polystyrene
beaker

50 cm³ of H₂O₂(aq),
concentration 0·88 mol l⁻¹

50 cm³ of
iron(III) nitrate
solution

As a result of the reaction, the temperature of the solution in the polystyrene beaker increased by 16 °C.

(i) What is the effect of the catalyst on the enthalpy change (ΔH) for the reaction?

1

(ii) Use the experimental data to calculate the enthalpy change, in $kJ\,mol^{-1}$, for the decomposition of hydrogen peroxide.

Show your working clearly.

3

(5)

Marks

15. (*a*) The graph shows how the freezing point changes with changing concentration for aqueous solutions of sodium chloride and ethane-1,2-diol.

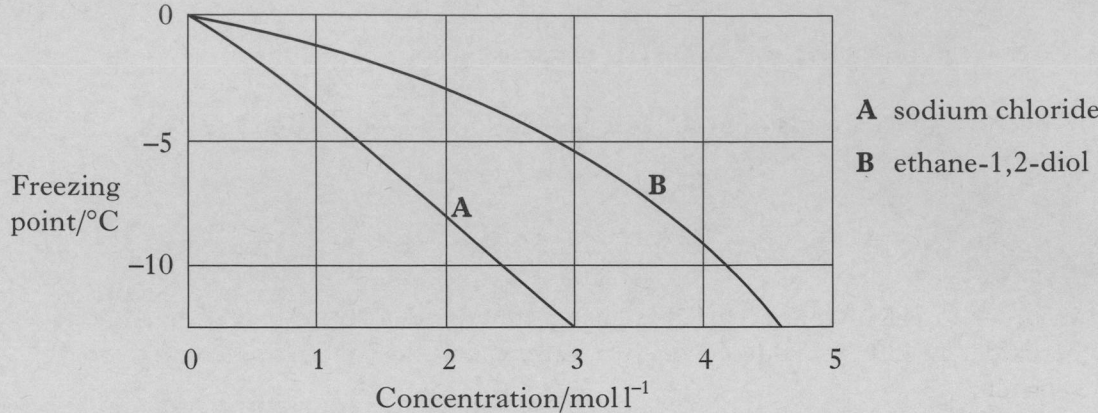

A sodium chloride

B ethane-1,2-diol

(i) Draw a structural formula for ethane-1,2-diol.

1

(ii) Ethane-1,2-diol solution is used as an antifreeze in car radiators, yet from the graph it would appear that sodium chloride solution is more efficient.

Suggest why sodium chloride solution is **not** used as an antifreeze.

1

(*b*) Boiling points can be used to compare the strengths of the intermolecular forces in alkanes with the strengths of the intermolecular forces in diols.

Name the alkane that should be used to make a valid comparison between the strength of its intermolecular forces and those in ethane-1,2-diol.

1

(3)

Marks

16. Aldehydes and ketones can take part in a reaction sometimes known as an aldol condensation.

The simplest aldol reaction involves two molecules of ethanal.

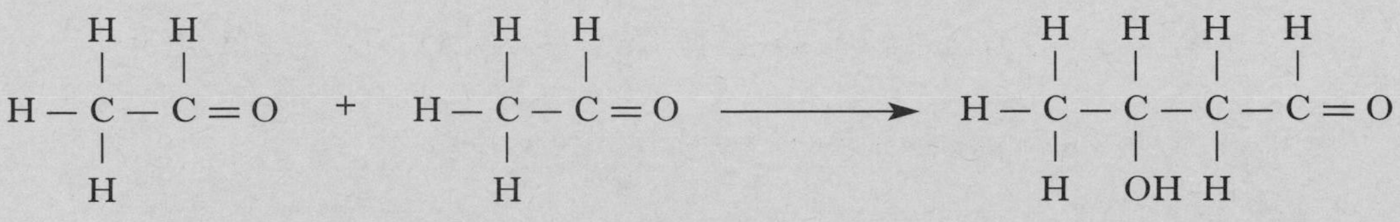

$$
\begin{array}{ccc}
\underset{\overset{|}{H}}{\overset{|}{H}}\,H-C-C=O & + & H-C-C=O \longrightarrow H-C-C-C-C=O
\end{array}
$$

In the reaction, the carbon atom next to the carbonyl functional group of one molecule forms a bond with the carbonyl carbon atom of the second molecule.

(a) Draw a structural formula for the product formed when propanone is used instead of ethanal in this type of reaction.

1

(b) Name an aldehyde that would **not** take part in an aldol condensation.

1

(c) Apart from the structure of the reactants, suggest what is unusual about applying the term "condensation" to this particular type of reaction.

1

(3)

[Turn over

Marks

17. Oxalic acid is found in rhubarb. The number of moles of oxalic acid in a carton of rhubarb juice can be found by titrating samples of the juice with a solution of potassium permanganate, a powerful oxidising agent.

The equation for the overall reaction is:

$$5(COOH)_2(aq) + 6H^+(aq) + 2MnO_4^-(aq) \rightarrow 2Mn^{2+}(aq) + 10CO_2(aq) + 8H_2O(\ell)$$

(*a*) Write the ion-electron equation for the reduction reaction.

1

(*b*) Why is an indicator **not** required to detect the end-point of the titration?

1

(*c*) In an investigation using a $500 \, cm^3$ carton of rhubarb juice, separate $25 \cdot 0 \, cm^3$ samples were measured out. Three samples were then titrated with $0 \cdot 040 \, mol \, l^{-1}$ potassium permanganate solution, giving the following results.

Titration	Volume of potassium permanganate solution used/cm^3
1	27·7
2	26·8
3	27·0

Average volume of potassium permanganate solution used = $26 \cdot 9 \, cm^3$.

(i) Why was the first titration result not included in calculating the average volume of potassium permanganate solution used?

1

Marks

17. *(c)* **(continued)**

(ii) Calculate the number of moles of oxalic acid in the $500 \, cm^3$ carton of rhubarb juice.

Show your working clearly.

2

(5)

[END OF QUESTION PAPER]

SPACE FOR ANSWERS

ADDITIONAL GRAPH FOR QUESTION 5(*b*)(i)

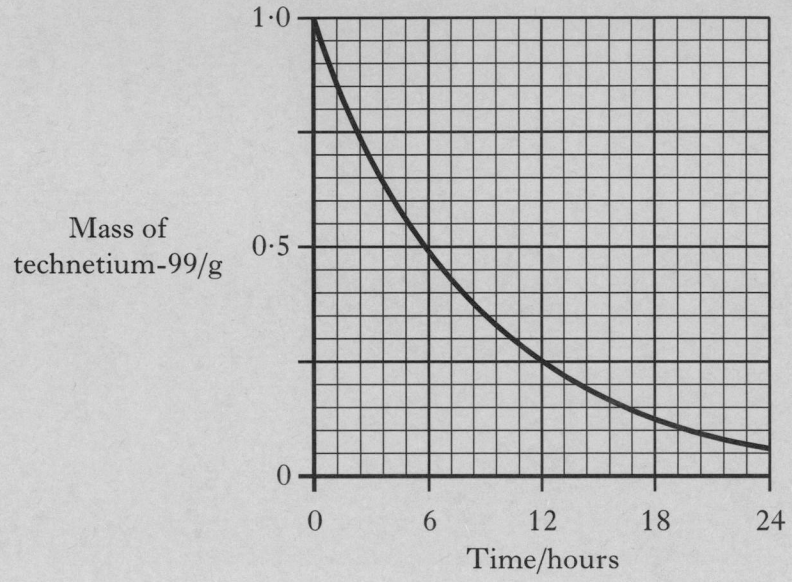

ADDITIONAL SPACE FOR ANSWERS

ADDITIONAL SPACE FOR ANSWERS

[BLANK PAGE]

[BLANK PAGE]

[BLANK PAGE]

[BLANK PAGE]

[BLANK PAGE]

[BLANK PAGE]